Collins

AQA GCSE 9-1
Combined
Science Trilogy
Higher

Ian Honeysett, Sam Holyman
and Jen Randall

Acknowledgements

The authors and publisher are grateful to the copyright holders for permission to use quoted materials and images.

Every effort has been made to trace copyright holders and obtain their permission for the use of copyright material. The authors and publisher will gladly receive information enabling them to rectify any error or omission in subsequent editions. All facts are correct at time of going to press.

Published by Collins
An imprint of HarperCollins*Publishers* Limited
1 London Bridge Street
London SE1 9GF

HarperCollins*Publishers*
Macken House
39/40 Mayor Street Upper
Dublin 1, D01 C9W8, Ireland

© HarperCollins*Publishers* Limited 2024

ISBN 978-0-00-867234-8

10 9 8 7 6 5 4 3 2 1

British Library Cataloguing in Publication Data.

A CIP record of this book is available from the British Library.

Authors: Ian Honeysett, Sam Holyman and Jen Randall
Publisher: Clare Souza
Commissioning: Richard Toms
Project Management and Editorial: Richard Toms and Katie Galloway
Inside Concept Design: Ian Wrigley
Layout: Rose & Thorn Creative Services Ltd
Cover Design: Sarah Duxbury
Production: Bethany Brohm
Printed in India by Multivista Global Pvt.Ltd.

MIX
Paper | Supporting responsible forestry
FSC
www.fsc.org
FSC™ C007454

This book contains FSC™ certified paper and other controlled sources to ensure responsible forest management.

For more information visit: www.harpercollins.co.uk/green

How to use this book

Each topic is presented on a two-page spread

Organise your knowledge with concise explanations and examples

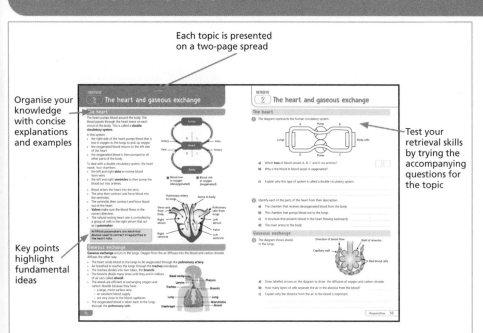

Key points highlight fundamental ideas

Test your retrieval skills by trying the accompanying questions for the topic

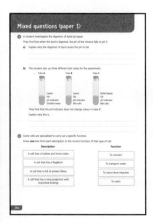

Mixed questions for each paper further test retrieval skills after all topics have been covered

Scientific and maths skills sections provide further knowledge and explanations of scientific and maths ideas and investigative skills

Answers are provided to all questions at the back of the book

Contents

7 Paper 2: Ecology

Chemistry

1 Paper 1: Atomic structure and the periodic table

2 Paper 1: Bonding, structure, and the properties of matter

3 Paper 1: Quantitative chemistry

4 Paper 1: Chemical changes

3 Paper 1: Particle model of matter

4 Paper 1: Atomic structure

5 Paper 2: Forces

6 Paper 2: Waves

7 Paper 2: Magnetism and electromagnetism

Animal cells

Inside cells are sub-cellular structures that carry out different functions.

A typical animal cell contains these sub-cellular structures:

- a **cell membrane** that controls which substances can enter or leave the cell
- a **nucleus** that contains the genetic material (DNA) and controls the reactions occurring in the cell
- a jelly-like substance called **cytoplasm**, in which many chemical reactions take place
- small structures in the cytoplasm called **mitochondria**, which are where respiration occurs
- **ribosomes**, where proteins are made.

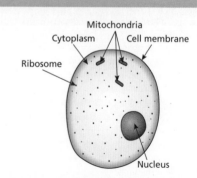

Plant cells

Plant cells have a cell membrane, cytoplasm, mitochondria and ribosomes – as animal cells do – but they may have other sub-cellular structures:

- a **cell wall** outside the cell membrane that is made of **cellulose** and supports the cell (algal cells also have cell walls)
- **chloroplasts** that absorb light and carry out all the reactions of photosynthesis to produce glucose
- a large permanent **vacuole** that contains a fluid called **cell sap**, which stores sugars and salts, and supports the cell.

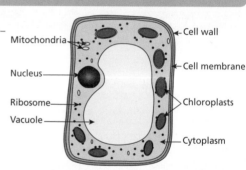

Not all plant cells have chloroplasts, for example, cells in the root will not have any because they do not get any sunlight.

Relating structure to function

The various sub-cellular structures are adapted to their roles in the cell:

- The cell membrane contains pores that let through some molecules and not others; this makes it **partially permeable**.
- The cell wall is made from many fibres of cellulose, which make it **permeable** but very strong.
- Chloroplasts have many stacks of membranes containing **chlorophyll** to trap as much light as possible.
- Mitochondria contain a membrane that is folded inwards to increase the surface area for respiration to occur.

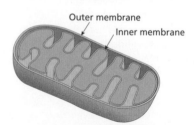

A chloroplast

Mitochondrion

1 Animal and plant cells

Animal cells

1 Draw lines to join each sub-cellular structure to its function.

Cell membrane	Respiration
Mitochondria	Controls what enters and leaves the cell
Nucleus	Protein production
Ribosomes	Contains the genetic material

Plant cells

2 Look at the diagram of a plant cell from a leaf.

Give the letters that label each of these structures.

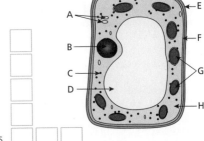

a) Cell membrane ☐

b) A structure made of cellulose ☐

c) A structure that contains cell sap ☐

d) The site of photosynthesis ☐

e) Three structures that are not found in animal cells ☐ ☐ ☐

3 How would this cell appear different if it was from a plant root?

....

Relating structure to function

4 Put a tick or a cross in each box in this table to show the features of mitochondria and chloroplasts.

	Mitochondria	Chloroplasts
Are the site of respiration		
Contain cell sap		
Are surrounded by a double membrane		
Contain chlorophyll		

5 How is the surface area of membranes increased inside mitochondria and inside chloroplasts?

....

....

....

1 Prokaryotic cells

Differences between prokaryotic and eukaryotic cells

There are two main types of cells: **prokaryotic** and **eukaryotic**.

There are a number of differences between the two types of cell:
- Prokaryotic cells are much smaller.
- The genetic material in prokaryotic cells is not enclosed in a nucleus.
- The genetic material in prokaryotic cells is a single DNA loop and there may be one or more small rings of DNA called **plasmids**.
- Prokaryotic cells do not have mitochondria or chloroplasts.

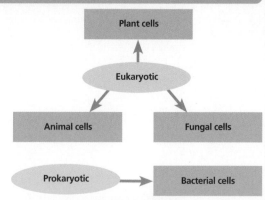

A typical bacterial cell

Bacterial cells can be different shapes. Some are round, some are rod-shaped and some are spiral, but they have features in common:
- The roles of mitochondria and chloroplasts in eukaryotic cells are taken over by the cytoplasm.
- There may be one or more **flagella**, which are tail-like structures that move the bacterium.
- Plasmids are present in the cytoplasm and they can be transferred from one cell to another, allowing the transfer of genes.

> Plasmids can contain genes for antibiotic resistance and so this can spread between bacteria.

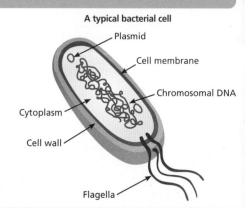

A typical bacterial cell

The size of cells

Cells are different sizes:
- a typical plant cell may be up to 0.1 mm and an animal cell up to 0.02 mm in diameter
- prokaryotic cells are smaller, often about 0.002 mm long.

To describe the size of cells and sub-cellular organelles, scientists usually use units that have different abbreviations.

Unit	Abbreviation	Number of these units in 1 metre
centimetre	cm	100
millimetre	mm	1000
micrometre	μm	1 000 000
nanometre	nm	1 000 000 000

> Make sure you can convert measurements from one unit to another.

Prokaryotic cells

Differences between prokaryotic and eukaryotic cells

1 Put a tick or a cross in each box of the table to show whether or not they are features of prokaryotic and eukaryotic cells.

	Prokaryotic cells	Eukaryotic cells
Contain mitochondria		
Contain DNA		
Contain cytoplasm		
Contain plasmids		

2 Describe the differences between chromosomes in prokaryotic cells and in eukaryotic cells.

A typical bacterial cell

3 Look at the diagram of a bacterium.

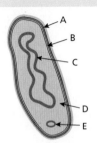

a) Give the letters that label each of these structures.

Cell wall ☐

Chromosome ☐

b) Describe how two bacteria of this species can exchange genes.

c) This species of bacterium cannot move on its own.

How can you tell this from the diagram?

The size of cells

4 The table shows some sizes of different cells and sub-cellular structures.

Cell / Structure	Size
Plant cell	0.1 mm
Animal cell	0.02 mm
Bacterium	0.002 mm
Nucleus	5 μm
Ribosome	25 nm

a) What is the size of a bacterium in micrometres (μm)?

b) How many times larger is a nucleus compared to a ribosome?

c) How many ribosomes could fit across the width of one animal cell?

1 Looking at cells

Using a light microscope to look at cells

The human eye can only see two objects as separate objects if they are more than 0.1 mm apart. Therefore, to study cells we need to use microscopes.

The ability to see two objects as separate objects is called **resolution**.

A light microscope has a greater resolution than the human eye and so can be used to see plant, animal and bacterial cells. Over time, the resolution of light microscopes has been improved but it is limited to 0.0002 mm by the wavelength of light.

See page 296 for details about how to use a light microscope to observe, draw and label a selection of plant and animal cells (Required Practical 1).

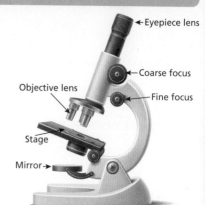

Eyepiece lens
Coarse focus
Objective lens
Fine focus
Stage
Mirror

Electron microscopes

In 1931, the first electron microscope was built. This allowed scientists to view smaller structures than is possible with the light microscope.

- The electron microscope has a higher resolution than the light microscope.
- The higher resolution allows a higher magnification to be used.
- Structures as small as 0.1 nanometres can be seen.

There are some disadvantages to using the electron microscope:

- the specimen has to have all the water removed so it cannot be alive
- it cannot detect colour.

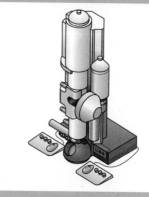

Calculating magnification

When a microscope is used to look at cells, scientists will often take photographs or produce drawings. These images are many times larger than the real cell or structure.

The **magnification** is how many times larger the image is than the real object.

$$\text{magnification} = \frac{\text{size of image}}{\text{size of real object}}$$

When using this formula, it is important to make sure both sizes are measured using the same unit.

Calculations involving sizes and magnifications often involve using **standard form** and **orders of magnitude**.

Standard form:

- is a system of writing numbers by using powers of 10
- is written in the form of $a \times 10^n$, where a is 1 or greater but less than 10 (e.g. 5000 is written as 5×10^3 and 0.005 is 5×10^{-3}).

Orders of magnitude:

- are a way of comparing the sizes of two structures
- are calculated by working out how many powers of 10 separate the sizes (e.g. 5000 is three orders of magnitude greater than 5).

1) Looking at cells

Using a light microscope to look at cells

1 What is meant by the term 'resolution'?

_____

2 What is the limit of resolution of the light microscope in micrometres? _____

Electron microscopes

3 When was the first electron microscope built?

_____

4 Put a tick or a cross in each box of this table to show the features of the light microscope and the electron microscope.

	Light microscope	Electron microscope
Produces coloured images		
Can be used to view bacteria		
Has a resolution greater than 0.0002 mm		
Can be used to view living cells		

Calculating magnification

5 The diagram shows the range of sizes for typical biological structures.

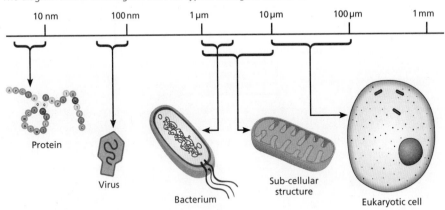

a) The actual width of the eukaryotic cell is 50 μm.

What is the magnification of this image? _____

b) What is the size of the largest eukaryotic cell?

Give your answer in metres and in standard form. _____

c) How many orders of magnitude separate the largest eukaryotic cell and the smallest bacterium?

_____

Chromosomes

- Each chromosome carries a large number of **genes**.
- The nucleus of a cell contains the genetic information. This is found on the chromosomes, which are made of molecules of DNA.
- Different genes control the development of different characteristics.
- In body cells, the chromosomes are normally found in pairs. One chromosome comes from each parent. Humans have 23 pairs of chromosomes in each body cell.

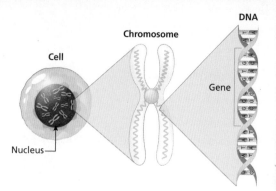

Humans have 23 pairs of chromosomes in each body cell but other species have different numbers, e.g. dogs have 39 pairs and horses have 32 pairs.

The cell cycle

Cells go through a series of changes involving growth and division.

This is called the **cell cycle**.

Before mitosis can occur again, the DNA replicates to form two copies of each chromosome.

At one stage of the cycle, the cell will divide into two by a process called **mitosis**.

The cell cycle

Each new cell that is produced grows and increases its numbers of sub-cellular structures such as ribosomes and mitochondria.

Mitosis

Once the chromosomes have replicated, mitosis can occur. One set of chromosomes is pulled to each end of the cell. Then the nucleus, the cytoplasm and the cell membrane divide to form two identical cells.

Mitosis

Parent cell with two pairs of chromosomes.

Each chromosome replicates (copies) itself.

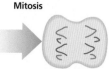
Chromosomes line up along the centre of the cell, divide and the copies move to opposite poles.

Each 'daughter' cell has the same number of chromosomes, and contains the same genes, as the parent cell.

Cell division by mitosis is important because it makes new cells for:
- growth and development of multicellular organisms
- repairing damaged tissues
- asexual reproduction.

Bacterial cells divide by binary fission. This is not the same as mitosis because bacteria do not have a nucleus.

1 Cell division

Chromosomes

1 What is a gene?

....

2 Put these structures in order of size, starting with the smallest.

chromosome   gene   cell   nucleus

....

3 Why do body cells have two copies of each chromosome?

....

The cell cycle

4 What happens in new cells after mitosis has occurred?

....

5 Explain why DNA needs to be copied before mitosis occurs.

....

....

Mitosis

6 The diagram shows a drawing of a cell in mitosis from a mosquito.

   a) How many chromosomes are in each mosquito body cell? ☐

   b) Describe what happens after the stage of mitosis shown in the diagram.

....

....

....

7 The pie chart shows the length of time for each stage in the cell cycle for liver cells. The whole cell cycle lasts 22 hours.

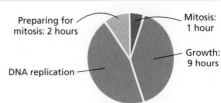

Preparing for mitosis: 2 hours

Mitosis: 1 hour

Growth: 9 hours

DNA replication

   a) Calculate how long DNA replication takes.

   b) Calculate the percentage of the cell cycle that is taken up by mitosis.

   c) Suggest why some types of cell have a shorter cell cycle than others.

....

(1) Stem cells

What are stem cells?

Most cells in the body of an adult organism can only carry out a particular job. When one of these cells divides by mitosis, it can only make new cells that carry out the same job.

However, some cells are different:
- Some cells are **undifferentiated**. This means that they can divide to make different types of cells. They are called **stem cells**.
- Stem cells from human embryos are called **embryonic stem cells** and can make all types of cells.
- Once a baby is born, it still has stem cells but they are called **adult stem cells**. They can only make certain types of cells.

> Although there are adult stem cells throughout the body, they are very difficult to find and isolate.

Stem cells in plants

In plants, stem cells are found in special areas called **meristems**:
- Meristem tissue is found in places such as root tips and shoot tips.
- Meristem cells are different to animal adult stem cells because they can produce any type of plant cell throughout the life of the plant.

Uses of stem cells

Stem cells can be used in humans and plants.

Potential uses for stem cells

In humans	In plants
Stem cells may be useful in replacing cells that are damaged or not working properly, such as in diabetes and paralysis.	Stem cells in the meristems can be used to produce whole new plants very quickly. These plants would be clones.
A cloned embryo of the patient may be made and this is used as a source of stem cells to treat a disorder; this is called **therapeutic cloning**.	Rare species of plants can be cloned to protect them from extinction.
Stem cells from the cloned embryo will not be rejected by the patient's body so they may be very useful in treating the patient.	Cloning allows large numbers of identical crop plants with special features (such as disease resistance) to be made.

> Some people are concerned about using stem cells from cloned embryos. They say that there may be risks such as the transfer of viruses. Other people may have ethical or religious objections.

1 Stem cells

What are stem cells?

1 How are stem cells different from normal body cells?

2 What is the difference between the function of embryonic stem cells and adult stem cells?

Stem cells in plants

3 What are meristems?

4 Name **one** place in a plant where meristem tissue is found.

5 It is easy to grow a new plant from a small piece of a leaf. However, it is not possible to grow a new animal from just a small piece of body tissue.

Explain this difference.

Uses of stem cells

6 Name a medical condition that could be treated using embryonic stem cells.

7 What is therapeutic cloning?

8 What is the advantage of treating a person using stem cells taken from an embryo cloned from that person?

9 A plant is resistant to a disease.

Explain the advantages of using stem cells from this plant to produce new plants.

(1) Diffusion

Diffusion

Many substances move into and out of cells across the cell membranes by **diffusion**.

- Diffusion is the net movement of particles from an area of higher concentration to an area of lower concentration.
- This happens due to the spreading out of the particles as they move randomly.
- There are many examples of diffusion in living organisms:
  - oxygen and carbon dioxide diffuse in lungs, gills and plant leaves
  - the waste product, urea, diffuses from cells into the blood for excretion by the kidneys
  - food molecules from the small intestine diffuse into the blood.

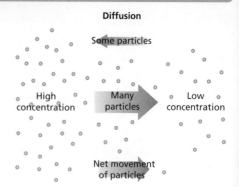

Diffusion

Some particles

High concentration   Many particles   Low concentration

Net movement of particles

Factors affecting diffusion

Factors which increase the rate of diffusion are:

- an increase in the difference in concentration of the molecules in the two areas – known as the **concentration gradient**
- an increase in temperature
- an increase in the surface area of the membrane separating the two areas.

> In diffusion, molecules move from high concentration to low concentration – this is called 'down the diffusion gradient'.

Speeding up diffusion

The surface area of a cell or organism compared to its volume is the **surface area to volume ratio**.

- Small or single-celled organisms have a large surface area to volume ratio. This allows enough molecules to diffuse into and out of the cell.
- In multicellular organisms, there is a smaller surface area to volume ratio. This means surfaces and organ systems must be specialised to speed up diffusion.

The small intestine and lungs in mammals, the gills in fish, and the roots and leaves in plants are all adapted for speeding up diffusion in many ways:

- they have a large surface area for diffusion to occur
- the surfaces are thin, to provide a short distance for the molecules to diffuse
- there is a rich blood supply in animals to maintain the concentration gradient
- ventilation occurs in animals to speed up gaseous exchange.

The effect of the size of an organism on its surface area to volume ratio

Organism A

Surface area = 6
Volume = 1
Surface area to volume ratio = 6

Organism B

Surface area = 24
Volume = 8
Surface area to volume ratio = 3

1 Diffusion

Diffusion

1 The diagram shows water molecules each side of a membrane.

Draw an arrow on the diagram to show the direction of the net diffusion of water.

2 Put one tick in each row in this table to show if the substance diffuses in or out of each part of the organism.

Part of the organism	Substance	Diffuses in	Diffuses out
Human liver cells	Urea		
Plant leaves in the light	Carbon dioxide		
Human muscle cells	Glucose		
Fish gills	Oxygen		

Factors affecting diffusion

3 Put a tick next to the change that will increase the rate of diffusion.

A change in temperature from 10°C to 20°C	
A decrease in a concentration gradient	
Smaller surface area of the membrane	

4 A person smells perfume when the lid of a bottle is taken off.

Explain why the person smells the perfume quicker on a warm day.

_____

Speeding up diffusion

5 The diagram shows three different cubes.

1 cm
Cube A

2 cm
Cube B

3 cm
Cube C

a) Complete the table for these three cubes.

Cube	Surface area in cm²	Volume in cm³	Surface area to volume ratio
A	6	1	6 : 1
B	24		
C		27	2 : 1

b) If these cubes represented living organisms, which organism, **A**, **B** or **C**, would need adaptations to speed up diffusion? Explain your answer.

_____

_____

1 Osmosis and active transport

Osmosis

Water may move across cell membranes by **osmosis**.

Osmosis is the diffusion of water from a dilute solution to a concentrated solution through a partially permeable membrane.

Water molecules are small enough to pass through the membrane but the dissolved solute molecules are too large.

Plant tissues such as potatoes can be used in the laboratory to measure the effect of sugar solutions on plant tissue:
- If potato pieces are left in pure water, they will take up water, increase in mass and get longer.
- If potato pieces are left in a concentrated sugar solution, they will lose water, decrease in mass and get shorter.
- If there is no change in mass or length, then the solution must be the same concentration as the contents of the potato cells.

How to investigate the effect of different concentrations of sugar or salt solutions on plant tissue is Required Practical 2 and is covered on page 297.

Osmosis

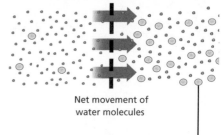

Dilute solution (high concentration of water) → Concentrated solution (low concentration of water)

Partially permeable membrane

Net movement of water molecules

These molecules are too large to pass through the membrane

Active transport

Active transport moves substances from a more dilute solution to a more concentrated solution.

This movement is against a diffusion gradient.

The energy for active transport comes from respiration.

Examples of active transport include:
- mineral ions being absorbed into plant root hairs from very dilute solutions in the soil
- sugar molecules being absorbed from lower concentrations in the gut into the blood, which has a higher sugar concentration.

Be ready to answer questions that compare diffusion, osmosis and active transport.

A cell absorbing ions by active transport

Root hair cell with high concentration of nitrate ions

Soil with low concentration of nitrate ions

Cell uses energy to pump ions against the concentration gradient

Osmosis and active transport

Osmosis

1 The diagram shows two sugar solutions each side of a membrane.

   a) Draw an arrow on the diagram to show the direction water moves by osmosis.

   b) Explain why the water molecules move through the membrane but the sugar molecules do not.

   _____

2 The cytoplasm of potato cells has a concentration equal to 36 g sugar / 100 ml.

   Explain what would happen to a cylinder of potato tissue left in each of these sugar solutions for an hour:

   a) 30 g sugar / 100 ml _____

   _____

   b) 36 g sugar / 100 ml _____

   _____

   c) 40 g sugar / 100 ml _____

   _____

3 Adding concentrated sugar solution to fruit makes jam and preserves the fruit. This is because it stops bacteria and fungi growing in the fruit.

   Suggest an explanation for this.

   _____

   _____

Active transport

4 Define 'active transport'.

   _____

   _____

5 Explain why it is necessary for active transport to be used to absorb sugar from the gut into the bloodstream.

   _____

   _____

6 Waterlogged soil does not contain much oxygen.

   Farmers try to make sure that the soil in the fields where they grow crops is not waterlogged when they add mineral fertilisers.

   Explain why.

   _____

   _____

   _____

(1) Specialised cells

Why cells specialise

Cells are the basic building blocks of all living organisms. The first cells in the embryo all look the same but, as an organism develops, cells become different types of cells.

- The process of changing into a particular type of cell is called **differentiation**.
- As a cell differentiates it may change shape to allow it to carry out a certain function. These cells are now **specialised cells**.
- Specialised cells are more efficient at their job.

> Although specialised cells are more efficient at their job, the disadvantage is that they lose the ability to do other jobs.

Specialised animal cells

Examples of specialised cells in animals are sperm cells, nerve cells and muscle cells.

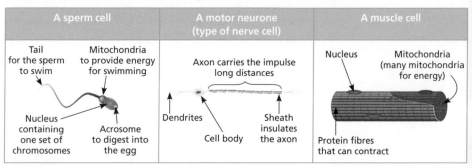

A sperm cell	A motor neurone (type of nerve cell)	A muscle cell

Specialised plant cells

Examples of specialised cells in plants are xylem cells, phloem cells and root hair cells.

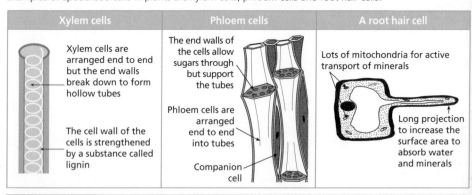

Xylem cells	Phloem cells	A root hair cell

> Most types of animal cell differentiate at an early stage, but many types of plant cells can differentiate throughout their life.

① Specialised cells

Why cells specialise

① Complete these sentences by writing the correct word in each gap.

Choose the words from this list.

different	differentiation	efficient	embryo	identical
	specialised	varied	zygote	

After fertilisation, a small ball of cells called an _____ is produced.

These cells all look _____.

After they divide many times, the cells become _____ for particular jobs.

This process is called _____.

The cells become more _____ at carrying out their function.

Specialised animal cells

② Complete this table, which gives information about specialised cells.

Cell type	Specialised feature	Function of specialised feature
		Enables the cell to swim to the egg
Muscle	Many mitochondria	
		Insulates the axon

③ Globozoospermia is a rare genetic condition in humans. Men with the condition make sperm that do not have an acrosome.

Suggest and explain an effect of this condition on reproduction.

_____

_____

Specialised plant cells

④ Add ticks and crosses to this table to show the presence (✓) or absence (✗) of the features in each of these cells.

Feature	Xylem cells	Phloem cells	Root hair cells
Walls containing lignin			
Holes in the ends of cells			
Projection from the side of the cell			

⑤ Explain why root hair cells need lots of mitochondria in order to take up minerals.

_____

_____

_____

② Tissues, organs and organ systems

Tissues

In most organisms, cells are arranged into **tissues**. A tissue is a group of cells with a similar structure and function. For example:

Muscle tissue	Glandular tissue	Epithelial tissue
Contracts so we can move	Produces substances such as enzymes and hormones	Covers organs

Although it can move around the body, blood is also a tissue.

Organs

Groups of tissues that perform specific functions are called **organs**. Each organ may contain several tissues. For example, the stomach is an organ that contains:

- muscle tissue, which contracts to churn the contents
- glandular tissue to produce digestive juices
- epithelial tissue to cover the outside and inside of the stomach.

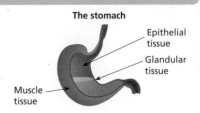

The stomach

Epithelial tissue

Glandular tissue

Muscle tissue

Organ systems

Organs are organised into **organ systems**.

Organ systems work together to form organisms. The **digestive system** is an example of an organ system in which several organs work together to digest and absorb food.

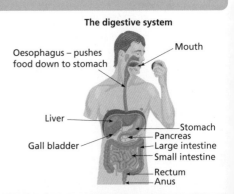

The digestive system

Oesophagus – pushes food down to stomach

Mouth

Liver

Gall bladder

Stomach
Pancreas
Large intestine
Small intestine
Rectum
Anus

Organisation in plants

Plants also have tissues, organs and organ systems. Some of these are shown in the table:

Tissues	Organs	Organ systems
Xylem	Leaf	Transport system
Phloem	Shoot	
Epidermis	Root	
Meristems		

② Tissues, organs and organ systems

Tissues

1. Write down a function of each of these human tissues:

muscle _____

glandular tissue _____

blood _____

Organs

2. Draw lines to join each stomach tissue with its function in the stomach.

Tissue	Function
Epithelial	Makes digestive juices
Glandular	Churns the food
Muscle	Covers the outside of the stomach

3. Look at the diagram of a section through an artery.

Explain why an artery is classed as an organ.

An artery

Layer of connective tissue

Thin layer of endothelial tissue

Layer containing muscle and elastic tissue

_____

_____

Organ systems

4. Name **three** organs found in the digestive system.

_____

Organisation in plants

5. Xylem, phloem, epidermis and meristems are all tissues found in plants.

   a) Which of these tissues are part of the transport system in plants?

   _____

   b) Which of these tissues transports water and minerals?

   _____

   c) Which of these tissues covers the surfaces of the plant?

   _____

② Enzymes

How enzymes work

Enzymes are biological **catalysts** that speed up chemical reactions in living organisms.

Enzymes have the following structure:
- they are all protein molecules
- they have a hole or groove in the molecule called the **active site**.

The **lock and key theory** is a model used to explain how enzymes work:
- the chemical that reacts is called the **substrate** (key)
- this fits into the active site on the enzyme (lock)
- the reaction takes place and the **products** leave the active site.

Each enzyme catalyses a specific reaction because the shape of the substrate (key) will only fit into an active site (lock).

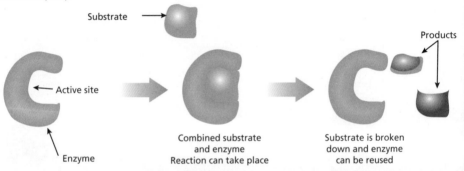

Substrate

Products

Active site

Enzyme

Combined substrate
and enzyme
Reaction can take place

Substrate is broken
down and enzyme
can be reused

Factors affecting enzymes

All enzymes work best at a specific temperature and pH.

- The 'best' temperature or pH is referred to as the **optimum**.
- At low temperatures, enzyme and substrate molecules move slowly and collide less often, so the reaction is slower.
- As the temperature increases, the collisions increase and so does the rate of reaction.
- At high temperature and extremes of pH, enzymes change shape; this is called **denaturing**.
- When the enzyme denatures, the substrate cannot fit into the active site, so the reaction slows down and stops.

> Do not make the mistake of saying that denaturing *kills* enzymes – enzymes are not living organisms.

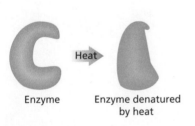

Enzyme

Enzyme denatured
by heat

Heat

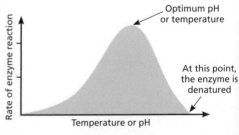

Rate of enzyme reaction

Optimum pH
or temperature

At this point,
the enzyme is
denatured

Temperature or pH

Required Practical 4 involves investigating the effect of pH on the rate of reaction of an enzyme and is covered on page 299.

2 Enzymes

How enzymes work

1 Complete these sentences about enzymes by writing the correct words in the gaps.

All enzymes are made of ....

They have a hole or groove on the molecule called the ....

Enzymes act as .... because they speed up chemical reactions in the body.

2 In the lock and key model of enzyme action, what does each of these represent?

a) the lock ....

b) the key ....

3 Explain why protease enzymes can break down protein but cannot break down starch.

....

....

Factors affecting enzymes

4 A student writes three sentences about enzymes.

1 *Enzymes are affected by temperature and the lowest temperature at which an enzyme can work is called the optimum.*

2 *As temperature increases, enzyme and substrate molecules will move more slowly.*

3 *The higher the temperature, the faster the reaction will happen.*

Explain the mistake that the student has made in each sentence.

1 ....

....

2 ....

....

3 ....

5 A student investigates the effect of temperature on the rate of reaction using two enzymes.

The graph shows the results for the first enzyme.

The second enzyme has:
- a higher optimum temperature
- a lower rate of reaction.

Draw a line on the graph to show the expected results for the second enzyme.

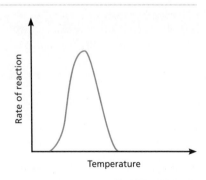

(2) Digestion

Enzymes and digestion

Food is broken down in the digestive system into smaller, soluble molecules so that it can be absorbed into the bloodstream. This is done by the enzymes **amylase**, **protease** and **lipase**.

- Amylase:
  - is a **carbohydrase** that digests starch
  - is produced in the salivary glands and pancreas
  - digests starch to maltose in the mouth and small intestine. (Other carbohydrases then digest maltose to glucose.)

- Protease:
  - digests proteins
  - is produced in the stomach and the pancreas
  - digests proteins into amino acids in the stomach and small intestine.

- Lipase:
  - digests lipids (fats and oils)
  - is produced in the pancreas and small intestine
  - produces fatty acids and glycerol in the small intestine.

> As well as making protease, the stomach also makes hydrochloric acid, which gives the optimum pH for the protease to work.

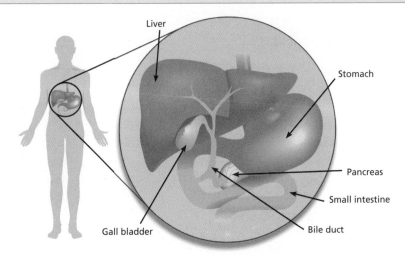

Using biological tests to identify starch, glucose, proteins and lipids is Required Practical 3 and is covered on page 298.

Bile and digestion

Bile is a liquid made in the liver and stored in the gall bladder.

- It is alkaline to neutralise hydrochloric acid from the stomach.
- It emulsifies fat to form small droplets, which increases the surface area.
- The alkaline conditions and large surface area increase the rate of fat breakdown by lipase.

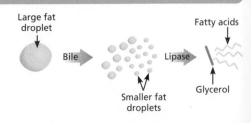

> Bile does not contain any enzymes, so it does not digest fat molecules; it just breaks up fat droplets.

② Digestion

Enzymes and digestion

1 Why does food need to be digested?

_____

_____

2 The diagrams show how different food molecules are digested in the human digestive system.

Complete the diagram by filling in the five gaps.

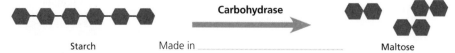

Starch   Made in _____ **Carbohydrase**   Maltose

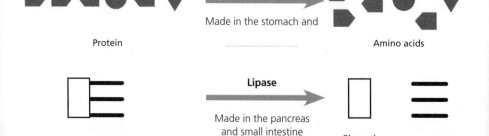

Protein   **Protease**   Made in the stomach and _____   Amino acids

Lipase   Made in the pancreas and small intestine   Glycerol

Bile and digestion

3 Complete these sentences about bile by writing suitable words in the gaps.

Bile is made in the _____.

It is stored in the gall bladder and passes to the small intestine through the

_____.

In the small intestine it _____ food, which has been made acidic in

the _____.

It also breaks down fat droplets into smaller droplets. This is called _____.

This increases fat digestion by _____.

4 Some people need to have their gall bladder removed.

Suggest why this might have an effect on fat digestion.

_____

_____

Blood

Blood is a tissue made of different types of cells in a liquid called **plasma**. Suspended in the plasma are:
- red blood cells
- white blood cells
- platelets.

The different components of blood have different functions:
- **Plasma** transports various chemical substances around the body, such as proteins, carbon dioxide, urea and glucose.
- **Red blood cells:**
  - contain a protein called **haemoglobin** which binds to oxygen to transport it from the lungs to the tissues
  - do not contain a nucleus, so that more haemoglobin can fit in
  - are very small so they can fit through the smallest blood vessels
  - are shaped like a biconcave disc, giving them a larger surface area so that oxygen can diffuse rapidly in and out.
- **White blood cells:**
  - help to protect the body against infection
  - can change shape so that they can squeeze out of the blood vessels into the tissues
  - can destroy microorganisms.
- **Platelets:**
  - are fragments of cells that collect at wounds and trigger blood clotting.

A red blood cell with a biconcave shape

A white blood cell destroying a group of microorganisms

There are different types of white blood cells, but they all have a nucleus and are larger than red blood cells.

Blood vessels

Blood passes around the body in three different types of **blood vessel**.
Each type of blood vessel is adapted to its function:

Arteries	Veins	Capillaries
• Take blood away from the heart to the organs. • Thick walls made of elastic and muscle fibres, to resist the high pressure of the blood.	• Take blood from the organs back to the heart. • Thinner walls and valves to prevent backflow as the pressure is lower.	• Join arteries to veins. • Narrow vessels with walls that are one cell thick, to allow substances to be exchanged with the tissues.

② Blood and blood vessels

Blood

1. The diagram shows the main parts of the blood and their functions.

Complete the diagram by writing words in the gaps.

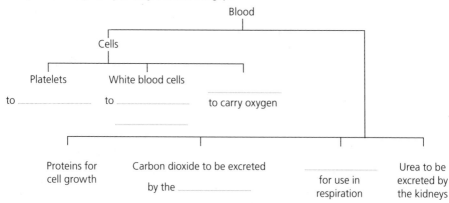

```
                                    Blood
              Cells

  Platelets      White blood cells
to ............      to ............        to carry oxygen

                          ............

  Proteins for   Carbon dioxide to be excreted    ............      Urea to be
  cell growth                                     for use in        excreted by
                    by the ............           respiration       the kidneys
```

2. The table shows the number of blood cells per mm^3 of blood in three different people.

Blood component	Person A	Person B	Person C
White blood cells	5000	8000	8000
Red blood cells	5 000 000	5 000 000	3 500 000
Platelets	3 000 000	1 500 000	3 000 000

   a) Person C gets tired very easily when running. Explain why.

   ....

   ....

   b) Compare the number of platelets in person B with the number of platelets in the other two people.

   What problems might the number of platelets in person B cause?

   ....

   ....

Blood vessels

3. Complete this table about blood vessels by putting a tick or a cross in each box.

	Artery	Vein	Capillary
Valves along its length			
Wall one cell thick			
Takes blood from organs to the heart			

The heart

The heart pumps blood around the body. The blood passes through the heart twice on each circuit of the body. This is called a **double circulatory system**.

In this system:
- the right side of the heart pumps blood that is low in oxygen to the lungs to pick up oxygen
- the oxygenated blood returns to the left side of the heart
- the oxygenated blood is then pumped to all other parts of the body.

To deal with a double circulatory system, the heart needs four chambers:
- the left and right **atria** to receive blood from veins
- the left and right **ventricles** to then pump the blood out into arteries.

- Blood enters the heart into the atria.
- The atria then contract and force blood into the ventricles.
- The ventricles then contract and force blood out of the heart.
- **Valves** make sure the blood flows in the correct direction.
- The natural resting heart rate is controlled by a group of cells in the right atrium that act as a **pacemaker**.

Artificial pacemakers are electrical devices used to correct irregularities in the heart rate.

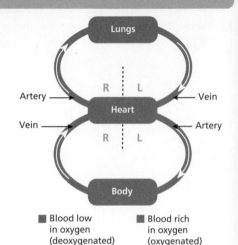

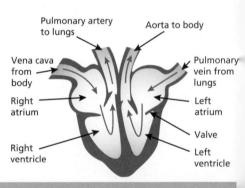

Gaseous exchange

Gaseous exchange occurs in the lungs. Oxygen from the air diffuses into the blood and carbon dioxide diffuses the other way.

- The heart sends blood to the lungs to be oxygenated through the **pulmonary artery**.
- Air breathed in reaches the lungs through the **trachea** (windpipe).
- The trachea divides into two tubes, the **bronchi**.
- The bronchi divide many times until they end in millions of air sacs called **alveoli**.
- The alveoli are efficient at exchanging oxygen and carbon dioxide because they have:
  - a large, moist surface area
  - an excellent blood supply
  - are very close to the blood capillaries.
- The oxygenated blood is taken back to the lungs through the **pulmonary vein**.

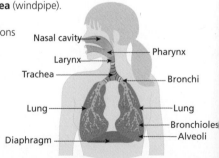

2) The heart and gaseous exchange

The heart

1 The diagram represents the human circulatory system.

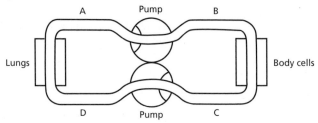

a) Which **two** of blood vessels A, B, C and D are arteries?

☐ ☐

b) Why is the blood in blood vessel A oxygenated?

_____

c) Explain why this type of system is called a double circulatory system.

_____

_____

2 Identify each of the parts of the heart from their description.

a) The chamber that receives deoxygenated blood from the body _____

b) The chamber that pumps blood out to the lungs _____

c) A structure that prevents blood in the heart flowing backwards _____

d) The main artery to the body _____

Gaseous exchange

3 The diagram shows alveoli in the lungs.

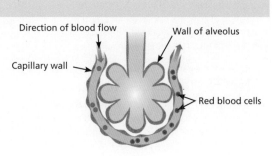

a) Draw labelled arrows on the diagram to show the diffusion of oxygen and carbon dioxide.

b) How many layers of cells separate the air in the alveolus from the blood? _____

c) Explain why the distance from the air to the blood is important.

_____

_____

Health and disease

Health is described as a state of physical and mental wellbeing.

A **disease** is caused by part of the body not working properly.

> Diseases are major causes of poor health, but other factors such as diet, stress and life situations may affect our physical and mental health.

Diseases can be divided into two main types:
- **Communicable** (infectious) **diseases** – can be spread between people, usually by a microorganism, and include tuberculosis, HIV and measles
- **non-communicable** (non-infectious) **diseases** – cannot be spread between people, and include cancer, diabetes, heart disease and inherited conditions.

There are many examples of different diseases interacting with each other:
- viruses infecting cells in the cervix can be the trigger for cancer
- diseases in the immune system mean that an individual is more likely to suffer from infectious diseases, e.g. people with HIV are more likely to get tuberculosis
- immune reactions caused by a pathogen can cause allergies, e.g. skin rashes and asthma
- if a person is physically ill, this can lead to depression and mental illness.

Risk factors

Non-communicable diseases are often caused by the interaction of a number of factors. These factors are called **risk factors** because they make it more likely for a person to develop the disease.

Risk factors can be:
- aspects of a person's lifestyle
- substances in the person's body or environment.

A graph can show that there is a pattern or correlation between a risk factor and getting a disease. However, this does not necessarily mean that the factor causes the disease.

> A pattern between two variables on a graph is called a correlation.

Scientists need to look for a **causal mechanism** to prove that a risk factor is involved. A causal mechanism has been found for some diseases and some risk factors; these are shown in the table:

Disease	Proven risk factors
Cardiovascular disease	Lack of exercise Smoking High levels of saturated fat in the diet
Type 2 diabetes	Obesity
Liver damage	Excess alcohol intake
Lung diseases including lung cancer	Smoking
Skin cancer	Ionising radiation, e.g. UV light

② Health and disease

Health and disease

1 Which of the following are communicable diseases / illnesses? Tick the correct answers.

Lung cancer ☐   Heart disease ☐

Covid-19 ☐   Chickenpox ☐

2 Draw lines to link the pairs of diseases that often interact with each other.

Virus infecting cells in the cervix	Tuberculosis
HIV / AIDS	Asthma
Immune reaction	Cancer

3 Suggest **one** reason why having a physical illness might affect a person's mental health.

Risk factors

4 Draw lines to link each disease with a risk factor for that disease.

Skin cancer	Obesity
Liver damage	Smoking
Type 2 diabetes	UV light
	Drinking excess alcohol

5 The graphs show information about smoking and annual death rates from cancer.

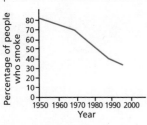

a) What overall correlation can be seen between smoking and the death rate from cancer?

b) Explain why there is a difference between the trend for smoking and the trend for cancer between 1950 and 1955.

② Heart disease and cancer

Heart disease

Coronary arteries lead off from the aorta just after it leaves the heart. These arteries spread over the surface of the heart muscle and supply it with oxygen and glucose for respiration.

In **coronary heart disease**, layers of fatty material build up inside the coronary arteries, narrowing them. This reduces the flow of blood through the coronary arteries, resulting in a lack of oxygen and glucose for the heart muscle. The artery may become completely blocked and this can cause heart failure.

Other types of heart disease can be caused by faulty heart valves. The valves may not open or close fully.

Coronary heart disease

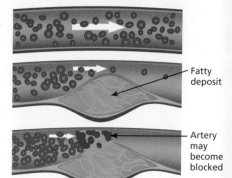

Fatty deposit

Artery may become blocked

Treatments for heart disease

Treatments for coronary heart disease include:
- inserting small tubes called **stents**, which keep the coronary arteries open
- taking drugs called **statins** to reduce the blood cholesterol level and slow down the rate of fatty material deposit.

Faulty heart valves can be replaced using:
- **biological valves** from an animal such as a pig – these work well and last for about 15 years
- **mechanical valves** which will last for life, but need drugs to prevent blood clotting.

For cases of heart failure:
- a donor heart, or heart and lungs, can be transplanted
- artificial hearts are occasionally used to keep patients alive whilst waiting for a heart transplant, or to allow the heart to rest in order to help recovery.

All treatments for heart disease have advantages and disadvantages that have to be considered when deciding on the best option for a patient.

Cancer

Cancer is caused by uncontrolled cell division. This can form masses of cells called **tumours**. There are two main types of tumours:
- **benign** tumours, which do not spread around the body
- **malignant** tumours, which spread to different parts of the body in the blood and can then form secondary tumours.

Cancer is a non-communicable disease, but scientists have identified risk factors for some types of cancers, for example:
- lifestyle risk factors, including smoking, obesity, common viruses and UV exposure
- genetic risk factors.

Cancers are usually treated by cutting out the tumour and giving drugs to prevent the cancer cells dividing (chemotherapy).

② Heart disease and cancer

Heart disease

1 Which blood vessel supplies blood to the coronary arteries?

_____

2 Explain why the coronary arteries are important for the correct functioning of the heart.

_____

_____

_____

3 Complete these sentences about heart disease.

Heart disease is caused by _____ deposits blocking the blood vessels that supply the heart muscle.

The heart muscle then does not get enough glucose or _____ for respiration.

This is called _____ heart disease.

4 Describe what would happen if a valve in the heart fails to close properly.

_____

_____

Treatments for heart disease

5 Some treatments for different heart problems are listed in the box.

| artificial hearts | chemotherapy drugs | human heart transplants |
| mechanical heart valves | pig heart valves | stents | statins |

a) Which treatment involves inserting a tube into the coronary artery?

_____

b) Which treatment slows down the build up of fatty deposits in the coronary arteries?

_____

Cancer

6 Explain why a malignant tumour is often more harmful than a benign tumour.

_____

_____

7 Why is it difficult to develop drugs that will kill cancer cells without damaging the body?

_____

_____

② Plant tissues

Examples of plant tissues

Like all tissues, plant tissues are groups of cells with a similar structure and function.
The table gives the function of different plant tissues:

Tissue	Function
Epidermal	Covers the outer surfaces of the plant for protection.
Palisade mesophyll	Has many chloroplasts and so is the main site of photosynthesis in the leaf.
Spongy mesophyll	Has air spaces between the cells so gases can diffuse through the leaf.
Xylem	Transports water and minerals through the plant. Also supports the plant.
Phloem	Transports dissolved food materials through the plant.
Meristem tissue	Found mainly at the tip of the roots and shoots, where it can produce new cells for growth.

Structure of a leaf

A plant leaf is an organ that is made of a number of different tissues. The structures of tissues in the leaf enable the plant to carry out its role of making food by photosynthesis.

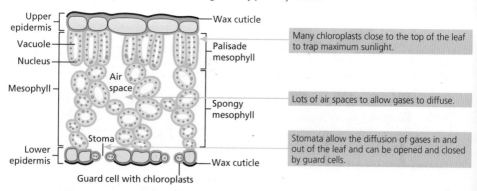

Many chloroplasts close to the top of the leaf to trap maximum sunlight.

Lots of air spaces to allow gases to diffuse.

Stomata allow the diffusion of gases in and out of the leaf and can be opened and closed by guard cells.

Guard cell with chloroplasts

The waxy cuticles are not tissues as they do not contain cells. They prevent the leaf losing too much water.

The lower epidermis has openings or pores called **stomata**. Stomata allow gases in and out of the leaf so that photosynthesis can happen. Surrounding the stomata are two guard cells:

- The role of guard cells is to open and close stomata.
- At night, the stomata are closed. This is because photosynthesis is not happening, so closing the stomata reduces water loss.
- Guard cells open stomata by taking up water and bending.
- Losing water makes the stomata close.

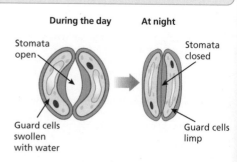

② Plant tissues

Examples of plant tissues

1 Identify each of these plant tissues from their description.

a) Made of cells joined end to end for transporting food materials. ....

b) Produces new cells for growth by mitosis. ....

c) Has cell walls strengthened with lignin to support the plant. ....

2 How is palisade mesophyll tissue adapted for its function?

....

....

Structure of a leaf

3 Write down the plant tissues in a leaf in order, starting from the tissue that is closest to the top surface of the leaf.

....

....

4 Write down the names of the tissues in a leaf that contain some chloroplasts.

....

....

5 The diagram shows sections through two different leaves.

Give **two** differences between the structure of the two leaves.

Leaf A   **Leaf B**

....

....

6 The diagram shows an image of the lower epidermis of a leaf seen with a light microscope.

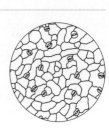

a) How many stomata are seen in this image? ....

b) The diameter of the image is 0.3 mm.

Calculate the area of the image.

Use the equation: area = πr^2 , where π = 3.14 and r = radius of the image

....

c) Use your answer to part b) to calculate how many stomata there are in 1 mm² of the lower epidermis.

....

Transpiration

Water enters the plant from the soil by osmosis through the root hair cells.

- This water contains dissolved minerals.
- The water and minerals are transported up the xylem vessels from the roots to the stems and leaves.

At the leaves, most of the water evaporates and diffuses out of the stomata. This loss of water from the leaves is called **transpiration**. Transpiration helps to draw the water up the xylem vessels from the roots.

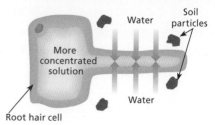

Less concentrated solution (dilute)

Water
Soil particles
More concentrated solution
Water
Root hair cell

Plants cannot stop water loss because the stomata need to be open for photosynthesis. Some plants that live in hot, dry places have ways to reduce transpiration.

Factors affecting transpiration

There are many factors that can affect the rate of transpiration:

- An increase in **temperature** will increase the rate because it will transfer more energy to the water to allow it to evaporate.
- An increase in **humidity** will decrease the rate as the air contains more water vapour and so the diffusion gradient is lower.
- Faster **air flow** will increase the rate as it will blow away water vapour allowing more to evaporate.
- Increased **light intensity** will increase the rate as it will cause the stomata to open.

Water uptake due to transpiration can be measured in a leafy shoot by using a piece of apparatus called a **potometer**.

Factors such as light intensity and air flow can be varied in the lab to see what effect this has on transpiration.

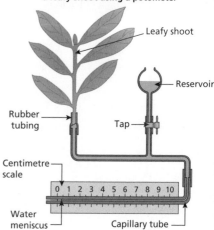

Measuring water uptake by a leafy shoot using a potometer

Leafy shoot
Reservoir
Rubber tubing
Tap
Centimetre scale
0 1 2 3 4 5 6 7 8 9 10
Water meniscus
Capillary tube

A potometer measures water uptake, which can be used to estimate transpiration. It is slightly lower as some of the water taken up is used in photosynthesis.

Translocation

Phloem tissue transports dissolved sugars from the leaves to the rest of the plant. This is called **translocation**. The main sugar translocated is sucrose. Phloem cells are adapted for this function (see page 22).

Transpiration

1 Explain why water can enter root hair cells by osmosis.

2 What is meant by the term 'transpiration'?

3 A student set up an experiment with three leafy shoots.

The layer of oil stops the water evaporating out of the test tube.

After 12 hours, the water level in each of the test tubes had dropped.

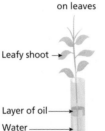
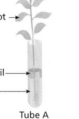
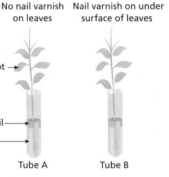

No nail varnish on leaves   Nail varnish on under surface of leaves   Nail varnish on upper surface of leaves

Leafy shoot →

Layer of oil —

Water —

Tube A   Tube B   Tube C

a) Explain why the water level in the test tubes dropped.

b) Which test tube, A, B or C, would have lost the least water?
Explain your answer.

Factors affecting transpiration

4 Put a tick in the box next to the conditions that will produce the fastest transpiration rates.

Warm and wet ☐   Dry and cold ☐

Windy and humid ☐   Windy and warm ☐

5 Give the reason why plants transpire faster in bright light conditions.

Translocation

6 Which is the main substance in a plant that is moved by translocation?

3 Viruses and disease

Pathogens

Pathogens are microorganisms that cause communicable (infectious) diseases.

Pathogens may infect plants or animals and can be spread by direct contact, by water or by air.

The spread of communicable diseases can be reduced by:
- simple hygiene measures, such as washing hands and sneezing into a handkerchief
- destroying **vectors**, which are organisms that pass on the pathogen without getting the disease
- isolating infected individuals so they cannot pass the pathogen on
- giving people at risk a **vaccination** (see page 48).

> During the Covid-19 pandemic, hygiene measures and isolating individuals were used to reduce the spread until a vaccine could be developed.

Viral diseases

Viruses reproduce rapidly in the cells of the body, causing damage to the cells.

Measles is a disease caused by a virus:
- The symptoms of measles are fever and a red skin rash.
- The virus is spread by breathing in droplets from sneezes and coughs.
- Although most people recover well from measles, it can be fatal if there are complications.
- Most people are vaccinated against measles as young children.

The measles vaccination was introduced into the UK in 1968 and dramatically reduced cases and deaths from measles.

HIV stands for human immunodeficiency virus:
- The virus is spread by sexual contact or exchange of body fluids, such as blood, which can happen when drug users share needles.
- If untreated, the virus enters the lymph nodes and attacks the body's immune cells.
- Late stage HIV, or AIDS, is when the body's immune system is damaged and cannot fight off other infections or cancers.

> There is sometimes a small health risk from medical treatments such as vaccinations, so it is important to weigh up whether the benefits are greater than the risks.

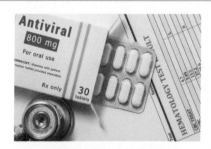

Antiviral drugs for treating HIV

> Taking antiviral drugs can now prevent AIDS occurring and so a person can live with HIV. However, in some countries these drugs are not always available.

An example of a plant viral pathogen is **tobacco mosaic virus** (**TMV**):
- It infects tobacco plants but also many other plants including tomatoes.
- It gives a distinctive 'mosaic' pattern of discolouration on the leaves.
- It affects the growth of the plant due to lack of photosynthesis.

Viruses and disease

Pathogens

1 What is the difference between a pathogen and a vector?

....

....

2 Explain why each of these measures can prevent the spread of disease:

Sneezing into a tissue: ....

....

Isolating individuals: ....

....

Viral diseases

3 Tick the boxes that correctly describe measles and / or AIDS.

	Measles	AIDS
Caused by a virus		
Spread by breathing in droplets		
Spread by sexual contact		
Is usually vaccinated against		
Can lead to death		

4 The graph shows the number of measles cases in the UK between 1940 and 2012.

a) Draw a smooth line of best fit on the graph to show the pattern in the number of measles cases between 1940 and 2012.

b) Suggest when the measles vaccination programme started in the UK. Give evidence from the graph.

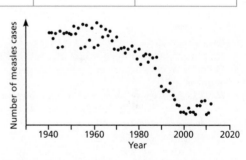

....

....

c) In 1998 a report was published claiming that there was a slight risk of side effects from the vaccination.

Explain the effect that this had on the data shown in the graph.

....

....

3 Bacteria, protists, fungi and disease

Bacterial diseases

Bacteria may damage cells directly or produce toxins (poisons) that damage tissues.

Salmonella is a type of food poisoning caused by bacteria:
- The bacteria are ingested in food, which may not have been cooked properly or which has been prepared in unhygienic conditions.
- The bacteria secrete toxins which cause fever, abdominal cramps, vomiting and diarrhoea.
- Chicken and eggs can contain the bacteria and so in the UK, chickens are vaccinated against salmonella to control the spread.

Gonorrhoea is a sexually transmitted disease (STD) and so is spread by sexual contact:
- The symptoms are a thick yellow or green discharge from the vagina or penis, and pain on urinating.
- As it is caused by a bacterium, it used to be easily treated with penicillin until many resistant strains appeared.
- The use of a barrier method of contraception, such as a condom, can stop the bacteria being passed on.

The bacteria that causes salmonella

Salmonella will not usually kill a healthy person, but it is more dangerous if a person is already unwell or is elderly.

A condom can protect against Gonorrhoea

Protists and disease

Protists are single-celled organisms but are **eukaryotic**, unlike bacteria.

Malaria is caused by a protist:
- The protist uses a particular type of mosquito as a vector.
- The protist is passed on when a person is bitten by the mosquito.
- Malaria causes severe fever which reoccurs and can be fatal.
- One of the main ways to stop the spread is to prevent people being bitten by mosquitoes. This is done by killing the mosquitos or by using nets to protect people from them.

Mosquitoes transmit malaria

Scientists have recently produced a vaccination against malaria. This was a difficult process as the pathogen changes form several times inside the human body.

Fungi and disease

An example of a fungal disease is **rose black spot**:
- It is spread when spores move from a rose plant to another plant by water or wind.
- Purple or black spots develop on the leaves, which often turn yellow and drop early.
- The growth of the plant is stunted as photosynthesis is reduced.
- It can be treated by using fungicides or by removing and destroying the affected leaves.

Bacteria, protists, fungi and disease

Bacterial diseases

1 What are toxins?

_____

_____

2 This is part of a leaflet giving advice on how to avoid getting salmonella.

Explain how each piece of advice will help prevent salmonella.

Wash hands and surfaces	Separate raw meats from other foods	Cook food to the right temperature	Refrigerate food properly
CLEAN	SEPARATE	COOK	CHILL

Clean: _____

_____

Separate: _____

_____

Cook: _____

_____

Chill: _____

_____

Protists and disease

3 Complete these sentences about malaria.

The pathogen that causes malaria is a _____.

Mosquitos act as _____ for this disease.

Sleeping under nets can give protection because it prevents mosquitos from _____.

Fungi and disease

4 The photograph shows a rose leaf infected by rose black spot.

a) What type of pathogen causes rose black spot?

_____

b) How can rose black spot be chemically treated?

_____

c) Explain why the disease effects the growth of the plant.

_____

_____

The body's defence against disease

Preventing entry of pathogens

The body has **non-specific defences** against disease. These defences work against all pathogens to try to stop them entering the body:

- The **skin** forms a barrier to prevent entry and makes an oily liquid called sebum which kills pathogens.
- The hairs in the **nose** trap particles that may contain pathogens.
- The **trachea** and **bronchi** are lined with cells that make mucus, which traps pathogens. Other cells have fine cilia which waft the mucus up to the mouth to be swallowed.
- The **stomach** has glands that produce hydrochloric acid, which kills pathogens.

> The body is particularly vulnerable to entry by pathogens in the natural openings, for example the eyes, ears, mouth and reproductive entrances, as exchange between the outside must happen here.

The immune system

If a pathogen enters the body, the **immune system** tries to destroy it.

White blood cells help to defend against pathogens by **phagocytosis** and by producing **antibodies** or **antitoxins**:

- **Phagocytosis** involves the pathogen being surrounded, engulfed and digested:

White blood cell			
Microorganisms invade the body.	The white blood cell finds the microorganisms and engulfs them.	The white blood cell ingests the microorganisms.	The microorganisms have been digested and destroyed.

- Special protein molecules called **antibodies** are produced, which attach to **antigen** molecules on the pathogen:

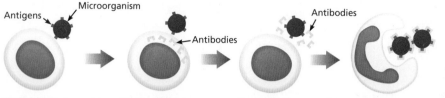

Antigens Microorganism		Antibodies	
Antigens are markers on the surface of the microorganism.	The white blood cells become sensitised to the antigens and produce antibodies. Antibodies	The antibodies lock onto the antigens.	This causes the microorganisms to clump together, so other white blood cells can digest them

- **Antitoxins** are produced, which are chemicals that neutralise the poisonous effects of toxins.

> The shape of an antibody is complementary to an antigen and so it can only attach to a specific antigen.

The body's defence against disease

Preventing entry of pathogens

1 The diagram shows four areas of the body (**A**, **B**, **C** and **D**) that help prevent entry of pathogens.

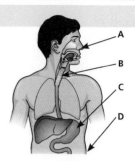

a) Which area has hairs to trap particles?

b) Which area kills pathogens using acid?

c) Explain how different types of cells in area **B** work together to protect the body from pathogens.

_____

_____

_____

2 Describe the function of sebum.

_____

The immune system

3 Draw lines to link each molecule to the correct description of the molecule.

Antibody		A molecule on the surface of a pathogen
Antigen		A molecule that neutralises poisonous substances
Antitoxin		A protein that attaches to the surface of antigens

4 The diagrams show stages in the destruction of a pathogen by a white blood cell. The stages are not in the correct order.

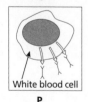

White blood cell

Pathogen

P   Q   R   S

a) Write the letters **P**, **Q**, **R** and **S** in the boxes to show the correct order of the stages.

b) Describe what is happening in stage **Q**.

_____

_____

Boosting immunity

Immunity

When a pathogen enters the body, white blood cells make antibodies to destroy the pathogen. However, this takes some time and the person can become ill (and possibly die) before the pathogens are destroyed.

Once the pathogen is destroyed:
- special white blood cells called **memory cells** stay in the bloodstream
- if the same pathogen re-enters the body, the memory cells respond more quickly to produce antibodies
- the pathogens are destroyed faster this time
- this prevents the person getting ill and this is called **immunity**.

> Immunity is not the same thing as resistance. Some people are naturally resistant to a disease, but immunity only happens after a person has had the disease (or a vaccination).

Vaccinations

A **vaccination** is a medical treatment that provides immunity to a disease. When a person has a vaccination:
- small quantities of dead or inactive forms of a pathogen are injected into the body
- the white blood cells are stimulated to produce antibodies and memory cells.

If the harmful pathogen invades the body again, the memory cells can produce antibodies to destroy the pathogen before the person gets ill.

If a large proportion of the population can be made immune to a pathogen, then the pathogen cannot spread very easily. This is called **herd immunity**.

> Sometimes the antigens on pathogens change and a new variant is produced. This means a new vaccination will need to be produced.

1 A weakened / dead strain of the microorganism is injected. Antigens on the modified microorganism's surface cause the white blood cells to produce specific antibodies.

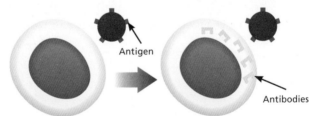

Antigen

Antibodies

2 The white blood cells that are capable of quickly producing the specific antibody remain in the bloodstream.

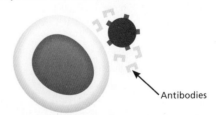

Antibodies

Boosting immunity

Immunity

1 Complete these sentences by writing words in the gaps.

When a pathogen enters the body, _____ make

_____ to destroy the pathogen.

Also, special cells called _____ are produced that can live for a long time.

If the same pathogens re-enter the body, the pathogens are destroyed faster.

This is because the person has _____ to the pathogen.

2 The graph shows antibody levels in a person after a pathogen first enters the body and after it enters for a second time.

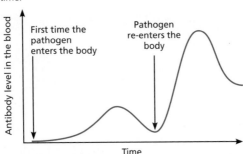

a) Describe **three** ways that the body's production of antibodies differs after the second time the pathogen enters, compared to the first time.

_____

_____

_____

b) Explain what causes the differences you described in part **a)**.

_____

_____

Vaccinations

3 What is contained in a vaccination that produces immunity?

_____

4 A person might have mild symptoms of a disease after having a vaccination.

Suggest why.

_____

_____

_____

Antibiotics and drug development

Antibiotics

Antibiotics, such as penicillin, are drugs that help to cure disease by killing bacteria inside the body.
- Doctors will prescribe certain antibiotics for certain types of bacteria.
- The use of antibiotics has greatly reduced deaths from bacterial diseases, but they cannot destroy viruses.

However, bacterial strains that are resistant to antibiotics are spreading (see page 80). **MRSA** is one example of a strain that is resistant to antibiotics.

To reduce the rate of development of antibiotic-resistant strains:
- doctors should not prescribe antibiotics unless they are really needed, and not for treating non-serious or viral infections
- patients must complete their course of antibiotics, so that all bacteria are killed and none survive to form resistant strains.

Developing new drugs

There is a constant demand for producing new drugs:
- New **painkillers** are developed to treat the symptoms of disease, however they do not kill pathogens.
- **Antiviral** drugs are needed that kill viruses without damaging the body's tissues.
- New antibiotics are needed, as resistant strains of bacteria develop.

Traditionally, new drugs were extracted from plants and microorganisms:
- **Digitalis** is a heart drug that originates from foxglove plants.
- **Aspirin** is a painkiller that originates from willow trees.
- **Penicillin** was discovered by Alexander Fleming from the *Penicillium* fungus.

Now, most new drugs are synthesised by chemists in the pharmaceutical industry. However, the starting point may still be a chemical extracted from a plant.

Willow tree

Foxglove

Testing new drugs

New medical drugs have to be trialled before being used, in order to find out:
- if they are safe (not toxic)
- if they work
- the optimum dose.

> **Preclinical testing is done in a laboratory using cells, tissues and live animals. Clinical trials use healthy volunteers and patients.**

Clinical trials on patients are usually **double blind trials**. This means that:
- some patients are given the drug; some patients are given a **placebo**, which does not contain the drug
- patients are allocated randomly to the two groups
- neither the doctors nor the patients know who has received a placebo and who has received the drug.

Antibiotics and drug development

Antibiotics

1 A student wrote these sentences about antibiotics and disease.

Antibiotics are used to kill bacteria and viruses on places such as door handles.
However, some bacteria such as MRSA cannot be killed as they are immune to many antibiotics.

Explain **three** mistakes the student has made in these sentences.

1 ....

2 ....

3 ....

2 A person is prescribed two weeks' supply of antibiotics, but they feel better after one week.

Why must they finish taking all the antibiotics?

....

....

Developing new drugs

3 Draw lines to connect each drug to where it is extracted from.

Drug	**Where it is extracted from**
Aspirin	Foxgloves
Digitalis	A fungus
Penicillin	Willow trees

4 Give a reason why it is important to constantly develop new antibiotics.

....

Testing new drugs

5 The pie chart shows the percentage of different animals used in animal experiments, such as drug testing.

a) Calculate the percentage of fish used in experiments.

....

b) Give **one** type of preclinical testing that does not use animals.

....

c) Very few monkeys are used in drug testing.

Suggest why using monkeys could be more useful than using mice but is rarely done.

....

....

....

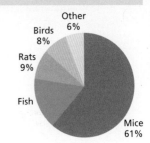

Other 6%
Birds 8%
Rats 9%
Fish
Mice 61%

4 Photosynthesis

The equation for photosynthesis

Plants use the Sun's energy in photosynthesis to make food.

The word equation for photosynthesis is:

$$\text{carbon dioxide + water} \xrightarrow{\text{light}} \text{glucose + oxygen}$$

The balanced symbol equation is:

$$6CO_2 + 6H_2O \xrightarrow{\text{light}} C_6H_{12}O_6 + 6O_2$$

To produce glucose molecules by photosynthesis, sunlight energy is required:

- this is because the reactions are **endothermic**
- the sunlight energy is trapped by the green chemical **chlorophyll**
- chlorophyll is found in chloroplasts.

> Endothermic reactions transfer in more energy or heat than they give out.

> Most of the chloroplasts in plants are in the palisade mesophyll layer of the leaves (see page 38).

Making other useful substances

The glucose produced in photosynthesis may be used by the plant during **respiration** to provide **energy**.

Glucose may also be changed into other products such as:

- insoluble starch, which is a store of energy in the stem, leaves or roots
- fat or oil, which is also an energy store, especially in some seeds
- cellulose, to strengthen cell walls
- amino acids to make proteins, which are used for growth and for enzymes
- chlorophyll for trapping the energy for photosynthesis.

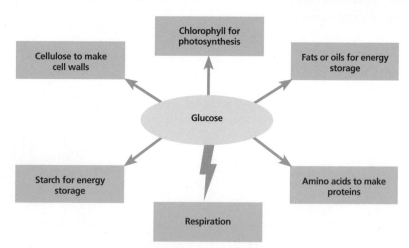

To make amino acids and proteins from glucose, plants need nitrate ions. They also need magnesium ions to make chlorophyll.

④ Photosynthesis

The equation for photosynthesis

1 Complete these sentences about photosynthesis by writing words in the gaps.

Photosynthesis needs energy, which plants trap from _____.

This energy allows plants to combine _____ and _____

together to make the carbohydrate _____.

The gas _____ is made as a by-product.

2 Complete the balanced symbol equation for photosynthesis.

_____ + _____ → $C_6H_{12}O_6$ + _____

3 Photosynthesis is described as endothermic. What is meant by the term 'endothermic'?

_____

_____

4 The picture shows a plant with variegated leaves. Variegated leaves are green in some parts and white in other parts. These plants are grown by gardeners because they look attractive.

Explain why they would have a disadvantage over other plants when growing together.

_____

_____

_____

Making other useful substances

5 Glucose can be converted into other substances.
Draw lines to join each substance to its function.

Substance	Function
Amino acids	Trapping sunlight
Fats	Making cell walls
Cellulose	Energy storage
Chlorophyll	Making proteins

6 Converting glucose into starch does not require any other substances, but converting glucose into amino acids requires nitrates. Explain why.

_____

_____

4 Factors affecting photosynthesis

Factors affecting the rate

There are several factors that may affect the rate of photosynthesis.

- **Temperature**: As the temperature increases, so does the rate of photosynthesis. As the temperature passes 45°C, the rate of photosynthesis drops to zero because the enzymes controlling photosynthesis have been denatured.

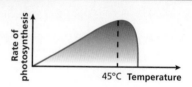

- **Carbon dioxide concentration**: As the concentration of CO_2 increases, so does the rate of photosynthesis. This is because CO_2 is needed in the reaction. After reaching a certain point, an increase in CO_2 has no further effect.
- **Chlorophyll concentration**: This does not vary in the short term but may change if plants are grown in soil without enough minerals to make chlorophyll.

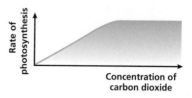

- **Light intensity**: As light intensity increases, so does the rate of photosynthesis. This is because more energy is provided for the reaction. After reaching a certain point, any increase in light has no further effect.

Investigating the effect of light intensity on the rate of photosynthesis is Required Practical 5 and is described on page 300.

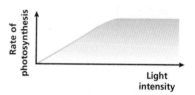

Limiting factors

At any moment, the factor that stops the photosynthesis reaction from going any faster is called the **limiting factor**:

- At high concentrations of carbon dioxide, the graph levels off because some other factor such as light or temperature becomes the limiting factor. CO_2 is no longer the limiting factor.
- At high light intensity, the graph levels off because some other factor such as carbon dioxide concentration or temperature becomes the limiting factor. Light intensity is no longer the limiting factor.

By looking at a graph, it can be possible to say what the limiting factor is at any point.

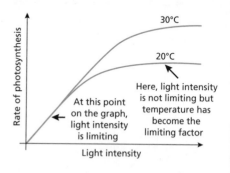

Greenhouses can be used to increase the rate of photosynthesis. By controlling lighting, temperature and carbon dioxide, farmers can increase the growth rate of their crops.

Factors affecting photosynthesis

Factors affecting the rate

1 Explain why light intensity has an effect on the rate of photosynthesis.

....

....

2 Why can the concentration of magnesium ions in the soil affect the rate of photosynthesis?

....

....

3 A student writes this prediction for the results of an experiment on photosynthesis.

As the temperature increases, the rate of photosynthesis increases until it levels off.
It levels off because the enzymes controlling the reaction start to die.

Explain **two** mistakes the student has made in their prediction.

1 ....

....

2 ....

....

Limiting factors

4 A gardener grows lettuces in the greenhouse shown in the diagram.

Explain why the lettuces grow better in the greenhouse than outside.

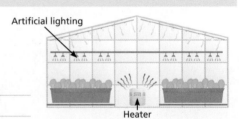
Artificial lighting

Heater

....

....

....

5 The graph shows the rate of photosynthesis of some plants grown in different conditions.

a) What is the limiting factor at each of these points?

A ....

C ....

b) Suggest a possible limiting factor at point **B**.

....

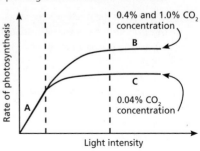

Rate of photosynthesis

0.4% and 1.0% CO_2 concentration

B

A

C

0.04% CO_2 concentration

Light intensity

Metabolism and respiration

Metabolism is the sum of all the chemical reactions in a cell or in the body.

- Metabolic reactions are controlled by **enzymes**.
- Some of these reactions need energy and some release energy.

Respiration is the main metabolic reaction that is **exothermic**. This means it transfers energy from molecules to the surroundings. The energy is released from **glucose** (a sugar).

> Energy is transferred between exothermic and endothermic reactions. Exothermic reactions transfer energy needed for endothermic reactions.

Organisms need energy released from respiration for:
- movement
- keeping warm
- chemical reactions to build larger molecules
- active transport
- the breakdown of excess proteins into urea for excretion.

The chemical reactions producing larger molecules include:
- the conversion of glucose to starch, glycogen and cellulose
- the formation of lipid molecules from a molecule of glycerol and three molecules of fatty acids
- the use of glucose and nitrate ions to form amino acids, which are used to synthesise proteins.

> Glucose is stored as starch in plants, but in animals it is stored mainly in the liver as glycogen.

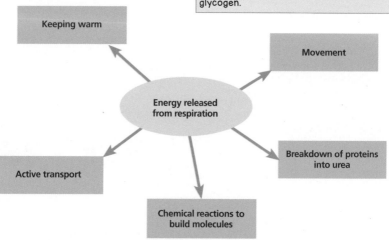

Aerobic respiration

Respiration in cells can be **aerobic** (uses oxygen) or **anaerobic** (without oxygen).

The equation representing aerobic respiration is the same in all organisms. The word equation for aerobic respiration is:

> glucose + oxygen → carbon dioxide + water

The balanced symbol equation is:

> $C_6H_{12}O_6 + 6O_2 \rightarrow 6CO_2 + 6H_2O$

> The equation for aerobic respiration is the reverse of the equation for photosynthesis.

Metabolism and aerobic respiration

Metabolism and respiration

1. Put a tick in the box next to the correct description of respiration.

An endothermic reaction
that releases energy. ☐

An exothermic reaction
that releases energy. ☐

An endothermic reaction
that requires energy. ☐

An exothermic reaction that
requires energy. ☐

2. Write down **two** types of processes in the body that require energy from respiration.

_____

_____

3. Draw lines to join each large molecule with the smaller molecule(s) that is used to make it.

Proteins

Fatty acids and glycerol

Lipids

Amino acids

Glycogen

Glucose

Aerobic respiration

4. Complete these sentences about respiration by writing words in the gaps.

The type of respiration that requires oxygen is called _____.

Water and _____ gas are given off as waste products.

5. A student set up this experiment to measure the rate of oxygen uptake in aerobic respiration by maggots.

The solution in the bottom of the test tube absorbs carbon dioxide.

Rubber tubing   Scale
0 1 2 3 4 5 6 7 8 9 10
Clip
Glass tubing
Capillary tube   Coloured liquid
Maggots
Gauze
Solution

a) Why does the coloured liquid move to the left when the maggots respire?

_____

_____

_____

b) Describe how the equipment can be used to measure the oxygen uptake by the maggots.

_____

_____

6. Complete the balanced symbol equation for aerobic respiration.

_____ + $6O_2$ → _____ + _____

4 Anaerobic respiration and exercise

Anaerobic respiration

Anaerobic respiration is respiration without the use of oxygen.

In animals, the process of anaerobic respiration is different to the process found in plants and yeast.

- In animals, the word equation is:

  glucose → lactic acid

- In plants and yeast, the word equation is:

  glucose → ethanol + carbon dioxide

- In plants and yeast, the symbol equation is:

  $$C_6H_{12}O_6 \rightarrow 2C_2H_5OH + 2CO_2$$

In anaerobic respiration, the glucose is not completely broken down. This means that much less energy is transferred from anaerobic respiration than from aerobic respiration.

Anaerobic respiration in yeast cells is called **fermentation** and is important in the manufacture of bread and different alcoholic drinks.

> Ethanol is the type of alcohol made by fermentation. The carbon dioxide made can make beer or wine fizzy and make bread rise.

Exercise

During exercise, the body needs more energy, so the rate of respiration needs to increase.
- The heart rate, breathing rate and breath volume increase.
- The muscles are supplied with more oxygen and glucose for aerobic respiration.

During long periods of vigorous activity, the muscles may not be supplied with enough oxygen. This results in:
- anaerobic respiration taking place in the muscles
- a build-up of **lactic acid** in the muscles (called an **oxygen debt**)
- the muscles hurting and not contracting efficiently due to the lactic acid.

Diffusion between a capillary and a working muscle cell

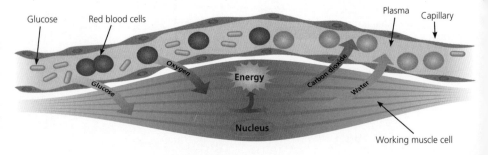

> The oxygen debt is the amount of extra oxygen the body needs after exercise to react with the lactic acid and remove it from the cells.

Once exercise is finished, the oxygen debt must be 'repaid'.
- Blood flowing through the muscles transports the lactic acid to the liver.
- It is then converted back into glucose.

4 Anaerobic respiration and exercise

Anaerobic respiration

1 Put a tick in the box next to the correct statement about anaerobic respiration.

It always requires glucose and oxygen.

In animals it produces lactic acid and carbon dioxide.

In plants it produces ethanol and water.

In yeast it produces carbon dioxide.

2 Students use this apparatus to investigate anaerobic respiration in yeast.

a) Explain why there are bubbles in the test tube.

_____

_____

_____

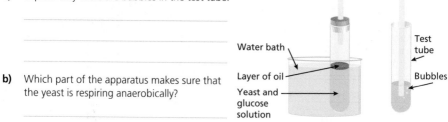

Water bath

Layer of oil

Yeast and glucose solution

Test tube

Bubbles

b) Which part of the apparatus makes sure that the yeast is respiring anaerobically?

_____

c) How could the students use the apparatus to measure the rate of anaerobic respiration?

_____

3 Complete the balanced symbol equation for anaerobic respiration in plants and yeast.

$$\text{_____} \rightarrow 2C_2H_5OH + \text{_____}$$

Exercise

4 Explain why it is important that the heart rate and breathing rate increase during exercise.

_____

_____

5 An athlete runs a race. After running for an hour, their muscles start to hurt. This is due to oxygen debt.

a) Which substance is making the muscles hurt? _____

b) Explain why this substance is produced in the muscles.

_____

_____

c) What happens to this substance after the athlete finishes the race?

_____

_____

(5) Homeostasis and control systems

Homeostasis

Homeostasis is the regulation of the internal conditions of a cell or organism in response to internal and external changes.

- Homeostasis keeps internal conditions in the body at optimum levels.
- This is important for enzyme action and all cell functions.

Homeostasis includes the control of:

- blood glucose concentration – this is important because glucose is the main substance used for respiration in the body
- body temperature – this is important because enzymes are affected by changes in temperature
- water levels – any change in the concentration of the blood can cause water to leave or enter cells by osmosis.

The control systems in homeostasis may involve:

- responses using nerves – the **nervous system** (see page 62)
- chemical responses using hormones – the **endocrine system** (see page 64).

Control of blood glucose concentration and control of water levels is by hormones. Control of body temperature is by nerves.

Control systems

The regulation of the internal conditions in the body is managed by control systems.

All control systems in the body include:

- **receptors**, which are cells that detect stimuli (changes in the conditions)
- **coordination** centres (such as the brain, spinal cord and pancreas) that receive and process information from the receptors
- **effectors** (muscles or glands), which instigate responses to bring the conditions back to optimum levels.

This type of control mechanism is called **negative feedback**.

This control mechanism is just like a thermostat in a house. It detects any change in temperature and then turns radiators on or off to bring the temperature back to normal.

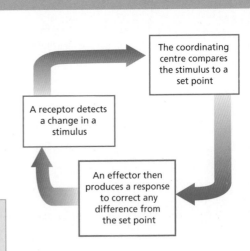

The coordinating centre compares the stimulus to a set point

A receptor detects a change in a stimulus

An effector then produces a response to correct any difference from the set point

Homeostasis and control systems

Homeostasis

1 Why is it so important to keep each of these factors constant in the body?

Blood glucose levels ....

....

Body temperature ....

....

2 Explain what would happen to red blood cells if the water level in the blood stayed too high.

....

....

....

3 Tick the factor that is controlled by nervous responses.

Blood glucose levels ☐

Body temperature ☐

Water levels in the blood ☐

Control systems

4 Draw lines to join each structure to its function in a control system.

Structure	Function
Coordinator	Bringing about a response
Effector	Detecting a stimulus
Receptor	Processing information

5 Blood pressure is controlled by negative feedback, as shown in the diagram.

Blood pressure

Heart muscle

Cells detecting the
stretching of blood vessels

Brain

In this negative feedback system, give the name of the:

Effector ....

Receptor ....

Coordinator ....

Organisation of the nervous system

The nervous system enables humans to react to their surroundings and to coordinate their behaviour. The two main parts of the nervous system are:

- the **central nervous system** (CNS), which is made up of the brain and spinal cord
- the nerves that spread throughout the body.

The CNS coordinates the response of effectors, i.e. muscles contracting or glands secreting hormones.

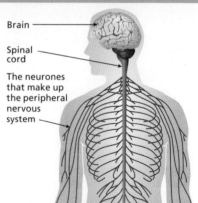

Brain

Spinal cord

The neurones that make up the peripheral nervous system

Synapses

Neurones are not directly connected to each other but communicate with each other by **synapses**:

- When an impulse reaches a synapse, a chemical is released.
- It diffuses across the gap between the two nerves.
- This then causes an impulse to be generated in the second nerve.

> Many drugs affect the body by either blocking or copying the action of neurotransmitters at synapses.

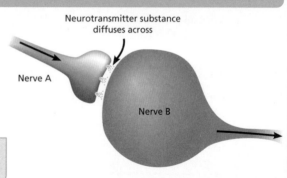

Neurotransmitter substance diffuses across

Nerve A

Nerve B

Reflex actions

Reflex actions are automatic and rapid, so they can protect the body; they do not involve the conscious part of the brain.

In a simple reflex action, such as a pain withdrawal reflex:

1. The pain stimulus is detected by receptors.
2. Impulses from the receptor pass along a sensory neurone to the CNS.
3. An impulse then passes through a relay neurone.
4. A motor neurone carries an impulse to the effector.
5. The effector, usually a muscle, withdraws the limb from the source of pain.

Required Practical 6 involves investigating human reaction times and is covered on page 301.

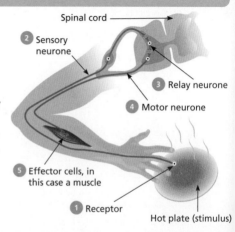

Spinal cord

2 Sensory neurone

3 Relay neurone

4 Motor neurone

5 Effector cells, in this case a muscle

1 Receptor

Hot plate (stimulus)

The nervous system and reflexes

Organisation of the nervous system

1 **a)** What do the letters CNS stand for?

....

b) Name the **two** parts of the CNS.

....

Synapses

2 **a)** How does a neurotransmitter cross a synapse?

....

b) Describe the function of a neurotransmitter at a synapse.

....

....

3 Painkillers are drugs that can reduce the pain sensed by the brain.

The diagram shows the shape of a neurotransmitter molecule found at synapses in neurones carrying pain signals to the brain.

It also shows the shape of a painkiller molecule.

Neurotransmitter molecule   Painkiller molecule

Explain how painkillers reduce the pain sensed by the brain.

....

....

Reflex actions

4 Put a tick next to the correct order of structures involved in a reflex action.

receptor → sensory neurone → relay neurone → motor neurone → effector ☐

effector → motor neurone → relay neurone → sensory neurone → receptor ☐

receptor → motor neurone → relay neurone → sensory neurone → effector ☐

5 Why is it important that a reflex, such as the pain withdrawal reflex, does not need to involve the brain?

....

6 The 'knee jerk' is caused by hitting a tendon in the knee. It causes the muscle that moves the leg to contract almost immediately. This protects the muscle from being stretched too much.

Give **two** reasons why this is a reflex.

....

....

⑤ The endocrine system

What are hormones?

Hormones are chemical messengers carried in the plasma of the blood to a target organ, where they produce an effect.

Hormones are produced and released by the endocrine system. **The endocrine system**:
- is composed of different glands
- secretes hormones directly into the bloodstream.

Hormones and nerves are alternative ways of passing messages around the body. Compared to nerve messages, hormone messages usually:
- produce a slower response
- produce effects that last longer
- affect a wider range of targets.

> Endocrine glands are sometimes called ductless glands because the hormones pass into the blood. Enzymes from glands like the salivary gland pass into tubes called ducts.

The main endocrine glands

There are different endocrine glands spread throughout the body. Here are some examples:

- The **pituitary gland** in the brain is a 'master gland' which secretes several hormones into the blood in response to body conditions. These hormones can act on other glands to stimulate other hormones to be released in order to bring about effects.
- The **pancreas** releases **insulin** and **glucagon**.
- The **ovaries** and **testes** produce the sex hormones:
  - **oestrogen** and **progesterone** from the ovaries
  - **testosterone** from the testes.

The **adrenal glands** produce **adrenaline**.

Adrenaline:
- is produced in times of fear or stress
- increases the heart rate and breathing rate
- boosts the delivery of oxygen and glucose to the brain and muscles, preparing the body for 'flight or fight'.

The **thyroid gland** produces **thyroxine**, which:
- increases the metabolic rate
- controls growth and development in young animals.

> The levels of thyroxine in the body are controlled by negative feedback.

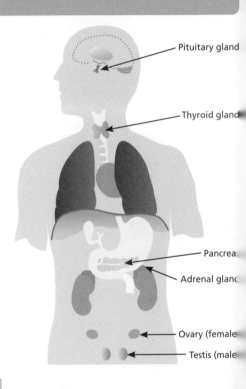

Pituitary gland

Thyroid gland

Pancreas

Adrenal gland

Ovary (female)

Testis (male)

5) The endocrine system

What are hormones?

1 Complete these sentences about hormones.

Hormones are released from _____ into the bloodstream.

They are carried around the body in the _____ of the blood.

The organ where they produce their effects is called the _____ organ.

2 Explain why hormone-secreting glands are called ductless glands.

_____

_____

3 Complete this table showing the differences between nervous and hormonal communication.

	Nervous	Hormonal
Type of signal used		chemical
Speed of response	fast	
Length of response		
Number of targets affected		numerous

The main endocrine glands

4 Identify the following hormones or endocrine glands in these descriptions.

a) The 'master' endocrine gland _____

b) The endocrine gland that produces progesterone _____

c) The male sex hormone _____

d) Endocrine glands found just above each kidney _____

e) An endocrine gland found in the neck _____

5 Adrenaline is sometimes called the 'fight or flight' hormone.

a) Explain how adrenaline helps to prepare the body for action.

_____

_____

_____

b) Most hormones have long-term effects. Explain why adrenaline is an exception.

_____

_____

_____

Homeostasis and response   65

⑤ Control of blood glucose

Hormonal control

Blood glucose concentration is monitored and controlled by the **pancreas**.

If the blood glucose concentration is too high:
- the pancreas releases more of the hormone insulin
- insulin causes glucose to move from the blood into the cells
- in liver and muscle cells, more glucose is converted to glycogen for storage.

If the blood glucose concentration is too low:
- the pancreas releases less insulin
- less glucose is converted to glycogen for storage.

The pancreas produces a second hormone called **glucagon**. This has opposite effects to insulin.

If the blood glucose concentration is too low:
- the pancreas releases less insulin but more glucagon
- glucagon stimulates glycogen to be converted to glucose.

If the blood glucose concentration is too high, less glucagon is released.

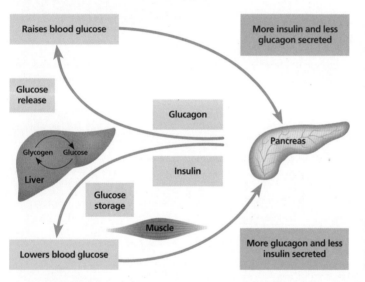

Raises blood glucose

Glucose release

Glucagon

Glycogen → Glucose

Liver

Insulin

Glucose storage

Muscle

Lowers blood glucose

More insulin and less glucagon secreted

Pancreas

More glucagon and less insulin secreted

Having two hormones working together means that blood glucose levels stay closer to the optimum.

It is important to spell the words glycogen and glucagon correctly, otherwise they can be confused.

Diabetes

Diabetes is a condition that means a person cannot properly control their blood glucose levels. There are two types of diabetes:

Type 1 diabetes	Type 2 diabetes
• Caused by the pancreas failing to produce sufficient insulin. • Characterised by uncontrolled high blood glucose levels. • Normally treated with insulin injections.	• Caused by the body cells no longer responding to insulin produced by the pancreas. • Obesity is a risk factor. • Treated with a carbohydrate-controlled diet and regular exercise.

⑤ Control of blood glucose

Hormonal control

1 What stimulates the release of insulin from the pancreas?

....

2 Which **two** organs contain the body's main store of glycogen?

....

....

3 Write down **two** ways that insulin decreases blood glucose levels.

....

....

4 Complete this table by adding ticks and crosses to show the features of insulin and glucagon.

	Insulin	Glucagon
Produced by the pancreas		
Released in response to an increase in blood glucose		
Causes more glucose to enter cells		
Stimulates the breakdown of glycogen		

Diabetes

5 Why can Type 2 diabetes not be treated with insulin injections?

....

6 The graph shows the blood glucose levels of a person with Type 2 diabetes and a person without Type 2 diabetes. The two people are given an insulin injection and have their blood glucose levels measured for two hours.

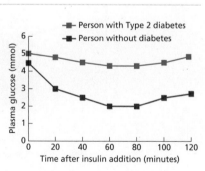

a) Compare the blood glucose levels of each person before the insulin is injected.

....

....

....

b) Explain why the insulin injection has a different effect on each person.

....

....

c) Predict what the graph would look like for a person with Type 1 diabetes.

....

....

5 Hormones and the menstrual cycle

The roles of hormones in reproduction

Hormones have many roles in controlling **human reproduction**.

- During puberty, the sex hormones cause secondary sexual characteristics to develop.
- After puberty in males, hormones stimulate the production of sperm.
- After puberty in females, hormones control the development and release of an egg, and changes in the uterus.

After puberty, men produce sperm continuously, but women have a monthly cycle of events called the **menstrual cycle**. The menstrual cycle involves:

- egg development
- release of an egg (**ovulation**) from the ovary
- and then the breakdown of the uterus lining (**menstruation**) if fertilisation has not occurred.

A number of hormones control reproduction:

- **Testosterone** is the main male sex hormone and is produced by the testes. It stimulates sperm production.
- **Oestrogen** is the main female sex hormone and is produced in the ovaries. It repairs the lining of the uterus after menstruation.
- **Progesterone** is released from the ovaries after ovulation. It maintains the lining of the uterus in the second half of the cycle.
- **Luteinising hormone** (LH) and **follicle stimulating hormone** (FSH) are both released from the pituitary gland. FSH stimulates egg development and LH stimulates ovulation.

Control of the menstrual cycle

The levels of the four main hormones change throughout the menstrual cycle. The hormones interact with each other to control the cycle:

- Oestrogen inhibits FSH release but stimulates LH release.
- Progesterone inhibits both FSH and LH release.
- FSH stimulates the ovaries to produce oestrogen.

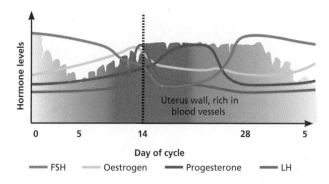

The average length of the menstrual cycle is 28 days but in some women it can be shorter or longer.

5 Hormones and the menstrual cycle

The roles of hormones in reproduction

1. The diagram shows the main hormone-producing glands in the body.

Give the letter that labels the gland where each of these hormones are made.

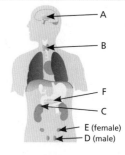

- A
- B
- F
- C
- E (female)
- D (male)

a) Oestrogen ☐   d) FSH ☐

b) Testosterone ☐   e) Progesterone ☐

c) LH ☐

2. Draw lines to join each hormone with its function.

FSH	Repairing the lining of the uterus
LH	Stimulating egg development
Oestrogen	Maintaining the lining of the uterus
Progesterone	Stimulating ovulation

Control of the menstrual cycle

3. The diagram shows the main events in the menstrual cycle.

a) Which hormone is responsible for the changes described between day 6 and day 10?

....

b) Oestrogen levels increase up to about day 13.

Explain how this leads to ovulation.

....

....

....

c) Progesterone levels drop at about 26 days onwards.

What effect does this have on the uterus?

....

....

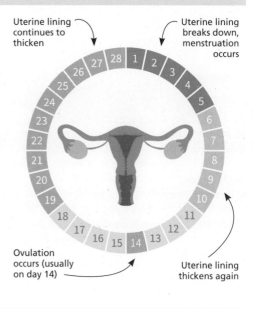

Uterine lining continues to thicken

Uterine lining breaks down, menstruation occurs

Ovulation occurs (usually on day 14)

Uterine lining thickens again

5 Controlling fertility

Reducing fertility

Fertility can be reduced by a variety of methods of **contraception**.

Hormonal methods of contraception include:
- **oral contraceptives** (the pill) that contain oestrogen and progesterone which inhibit FSH production so that no eggs are released
- **injection**, **implant** or **skin patch** of slow-release progesterone to stop the release of eggs for a number of months or years.

Non-hormonal methods of contraception include:
- barrier methods such as **condoms** and **diaphragms** so the sperm cannot reach an egg
- **intrauterine devices** which prevent an embryo implanting in the uterus
- **spermicidal creams**, which kill or disable sperm
- **timing intercourse** to avoid when an egg may be in the oviduct
- surgical methods of male and female **sterilisation**, such as cutting the sperm ducts or tying the fallopian tubes.

Some methods of contraception are more effective than others at preventing pregnancy.

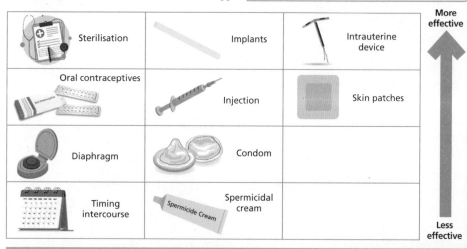

More effective

Sterilisation	Implants	Intrauterine device
Oral contraceptives	Injection	Skin patches
Diaphragm	Condom	
Timing intercourse	Spermicidal cream	

Less effective

Increasing fertility

Different methods can be used to try to help a woman get pregnant.

Fertility drugs:
- contain FSH and LH to stimulate eggs to mature and be released
- may be used if a woman's own hormone levels are too low.

In Vitro Fertilisation (IVF) treatment, which involves:
- giving a woman FSH and LH to stimulate the growth of multiple eggs
- collecting the eggs from the woman
- fertilising the eggs with sperm from the father in the laboratory
- inserting one or two embryos into the woman's uterus (womb).

Fertility treatment gives a woman the chance to have a baby of her own but:
- can be very emotionally and physically stressful
- the success rates are not high
- it can lead to multiple births, which are a risk to both the babies and the mother.

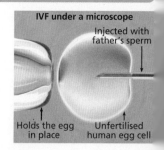

IVF under a microscope

Injected with father's sperm

Holds the egg in place

Unfertilised human egg cell

(5) Controlling fertility

Reducing fertility

1 Complete the table by adding **one** tick on each row to show the type of contraceptive method.

	Hormonal	Non-hormonal
Oral contraceptive		
Condom		
Skin patch		
Diaphragm		
Sterilisation		

2 Draw lines to join each contraceptive method to the description of how it works.

Method	How it works
Spermicidal cream	Stops sperm entering the vagina after release
Condom	Prevents an egg leaving the oviduct
Contraceptive pill	Kills sperm
Female sterilisation	Inhibits egg release

3 Discuss the advantages and disadvantages of using sterilisation as a method of contraception.

Increasing fertility

4 Some women have difficulty getting pregnant because they have blocked oviducts.

Explain why IVF can help these women to get pregnant.

5 During IVF treatment, 10 to 15 eggs are collected from a woman's ovaries.

Suggest **two** reasons why so many eggs are collected.

Reproduction and meiosis

Asexual reproduction

Asexual reproduction occurs in many different types of organism. Asexual reproduction involves:
- only one parent
- no fusion of gametes, so no mixing of genetic information
- the production of genetically identical offspring (**clones**)
- mitosis but not meiosis.

Many plants reproduce asexually in different ways, for example:
- strawberry plants send out long shoots called **runners** that touch the ground and grow a new plant
- daffodil plants grow lots of smaller bulbs that can grow into new plants.

Daffodils

Strawberry plant

Bulb

Runner

Sexual reproduction

Sexual reproduction involves the joining (fusion) of male and female **gametes**. In animals:
- the male gametes are sperm and the female gametes are egg cells
- sperm are made in the testes and egg cells are made in the ovaries.

In flowering plants:
- the male gamete is in the pollen and the egg cells are in the ovule inside the flower
- the pollen has to be transferred to the

female part of the flower
- this allows the male and female gametes to fertilise.

Because sexual reproduction involves fertilisation of male and female gametes, the offspring have genetic material from both parents.

> Children will have a mixture of genetic material from both parents. Siblings will therefore often look a little like each other and also resemble both parents.

Meiosis

The formation of gametes involves **meiosis**.

When a cell divides by meiosis:
- copies of the genetic information are made
- the cell divides twice to form four gametes, each with a single set of chromosomes
- all gametes are genetically different from each other.

Meiosis is important because it halves the number of chromosomes in gametes so that fertilisation can restore the full number of chromosomes.

Sperm		Egg		Fertilised egg cell
	+		=	
23 chromosomes	+	23 chromosomes	=	46 chromosomes (23 pairs) – half from mother (egg) and half from father (sperm)

Meiosis

Cell with two pairs of chromosomes (diploid cell).	Each chromosome replicates itself.	Chromosomes part company and move to opposite poles.	Cell divides for the first time.	Copies now separate and the second cell division takes place.	Four haploid cells (gametes), each with half the number of chromosomes of the parent cell.

Reproduction and meiosis

Asexual reproduction

1 The diagram shows a strawberry plant.

a) Explain how this strawberry plant can reproduce asexually.

....

....

....

....

b) What evidence is there on the diagram that this strawberry plant can reproduce sexually as well as asexually?

....

Sexual reproduction

2 Explain why children often have features that resemble both parents.

....

....

Meiosis

3 Write **true** or **false** next to each of these statements about meiosis.

Meiosis produces two cells from one cell. ....

Meiosis is used to produce body cells. ....

Meiosis produces genetically different cells. ....

Each cell produced by meiosis has one chromosome from each pair. ....

4 The drawing shows a cell before meiosis.

Which of the cells, A, B, C or D, would be produced by meiosis?

Before   **A**   **B**   **C**   **D**

5 An elephant has 56 chromosomes in a skin cell.
Complete this table to show the number of chromosomes found in different elephant cells.

Type of cell	Number of chromosomes
Elephant skin cell	56
Elephant sperm cell	
Elephant cheek cell	
Elephant egg cell	

6 DNA and the genome

The genome

The genetic material is in the nucleus of a cell and is made of a chemical called **DNA**. DNA is a polymer because it is made up of repeating units called nucleotides. The nucleotides are joined together to form long strands which are twisted to form a double helix.

The DNA is contained in structures called **chromosomes** containing **genes**.

Each **gene**:

- is a small section of DNA on a chromosome
- codes for a particular sequence of amino acids to make a specific protein.

The **genome** of an organism is the entire genetic material of that organism.

The whole human genome has now been studied and this may have some important uses in the future. It may:

- allow doctors to search for genes linked to different types of disorders
- help scientists to understand the causes of, and how to treat, inherited disorders
- be useful in working out how humans moved about the Earth in the past.

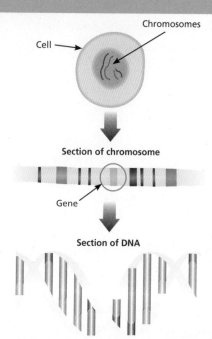

Section of chromosome

Gene

Section of DNA

A person now has the choice to have their genome tested to see how likely it is that they may get certain disorders. This may be a difficult decision to make.

Genes and alleles

Some characteristics are controlled by a single gene, such as:

- fur colour in mice
- attached or free ear lobes in humans.

Each gene may have different forms called **alleles**. For example, the gene for the attachment of ear lobes has two alleles, either attached or free.

- The alleles that are present in a person are called the **genotype**.
- How the alleles are expressed (what characteristic appears) is called the **phenotype**.

Alleles can either be **dominant** or **recessive**:

- A dominant allele is always expressed, even if only one copy is present.
- A recessive allele is only expressed if two copies are present (therefore, no dominant allele is present).

An individual always has two alleles for each gene:

- one allele for each gene comes from the mother
- one allele for each gene comes from the father.

The two alleles present can be the same or different:

- If they are the same, the person is **homozygous**.
- If the alleles are different, they are **heterozygous** for that gene.

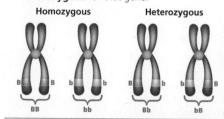

Many characteristics are not controlled by a single gene but are a result of many genes interacting.

6 DNA and the genome

The genome

1 What is a gene?

_____

_____

2 Write down these structures in order of size, starting with the smallest.

| cell | chromosome | gene | nucleus |

Smallest: _____

_____

_____

Largest: _____

3 A person can now have their genome tested.

The table shows the aims of four different types of test.

	Aim of test
P	To find DNA that has come from a cancer tumour.
Q	To find the best drug to treat a genetic disorder.
R	To identify genes that might affect the chance of developing a disease.
S	To find out about where a person's relatives came from.

a) Which test(s) from **P**, **Q**, **R** or **S** is / are useful if somebody already has a disease or disorder?

_____

b) Test **R** might tell a person that they have a 10% chance of developing a disease such as heart disease in the next five years.

Suggest why information like this can be useful for the person to know.

_____

_____

Genes and alleles

4 Complete the gaps in these sentences, using words from this list.

| chromosome | four | gene | genotype | one | phenotype | two |

An allele is an alternative form of a _____ .

In each body cell, there are _____ alleles for each gene.

The alleles that are present in a person is called the _____ .

The characteristic that is produced due to the alleles is called the _____ .

Genetic crosses and sex determination

Genetic crosses

Most characteristics are controlled by several genes working together, but if only one pair of alleles is involved, it is referred to as **monohybrid inheritance**.

Genetic diagrams or **Punnett squares** can be used to predict the outcome of a monohybrid cross. These diagrams use:

- a capital letter for dominant alleles
- a lower-case letter for recessive alleles.

So, for ear lobes, the free allele is dominant, so **E** can be used; for the unattached allele (recessive) **e** can be used. The Punnett squares show the possible outcomes of two different crosses.

> The offspring predicted by using a Punnett square is only a probability. The actual numbers might vary due to chance.

Some human disorders are inherited. These disorders are caused by the inheritance of certain alleles:

- **Polydactyly** (having extra fingers or toes) is caused by a dominant allele.
- **Cystic fibrosis** (a disorder of cell membranes) is caused by a recessive allele.

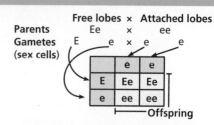

Each offspring will have a 1 in 2 chance of having the phenotype for free lobes (because the dominant allele is present in half the crosses).

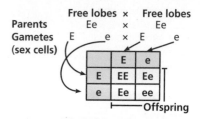

Each offspring has a 3 in 4 chance of having the phenotype for free lobes (because the dominant allele is present in three out of four crosses).

Sex determination

Only one pair out of the 23 pairs of chromosomes in the human body carries the genes that determine sex. These are called the **sex chromosomes**.

- In females, the sex chromosomes are identical and are called the X chromosomes.
- In males, one sex chromosome (Y) is much shorter than the other (X).

As with other pairs of chromosomes, offspring inherit:

- one sex chromosome from the mother (always an X chromosome)
- one sex chromosome from the father (either an X or a Y chromosome).

So, the sex of the offspring is decided by whether the ovum (egg) is fertilised by an X-carrying sperm or a Y-carrying sperm.

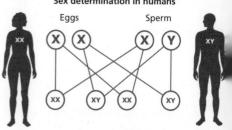

Sex determination in humans

Using a Punnett square to cross XY with XX shows that the ratio of males to females in the offspring should be 1 : 1.

Genetic crosses and sex determination

Genetic crosses

1 Cystic fibrosis is a condition that causes thick fluid to build up in the lungs.
It is controlled by one gene. The allele for cystic fibrosis is f.

a) Why is the inheritance of cystic fibrosis described as monohybrid?

_____

b) From the information in the question, how can you tell that cystic fibrosis is caused by a recessive allele?

_____

c) Two parents have a child who has cystic fibrosis. Neither of the parents has the condition.

Complete the Punnett square to predict the probability that their next child will have cystic fibrosis.

		Mother	
		F	f
Father	F		
	f		

Probability = _____

2 Fur colour in mice is controlled by a single gene. The allele for grey fur (G) is dominant over the allele for white fur (g).
Two grey mice are mated twice and produce in total twenty mice.
Seventeen of the mice are grey and three are white.

a) What is the genotype of the two parent mice?

_____

b) What numbers of grey and white mice in the twenty offspring would be predicted to be produced by the two matings?

_____

c) How would you account for the difference between the observed and the predicted numbers?

_____

_____

Sex determination

3 Put a tick next to the correct statement concerning sex determination.

An egg always contains two X chromosomes. ☐

An egg always contains one X chromosome. ☐

A sperm always contains a Y chromosome. ☐

A sperm always contains an X and a Y chromosome. ☐

Variation and natural selection

Variation

Differences in the characteristics of individuals in a population are called **variation**.

This variation may be due to:
- differences in the genes they have inherited (**genetic causes**)
- differences in the conditions in which they have developed (**environmental causes**)
- a combination of differences in genes and environment.

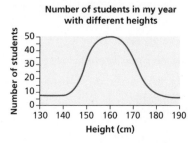

Number of students in my year with different heights

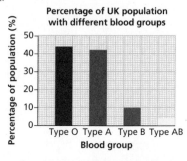

Percentage of UK population with different blood groups

ABO blood groups are controlled by just genes, but height is a combination of genes and environment. That is why there is a wide range of possible heights.

Sexual reproduction combines alleles in different combinations to produce variation, but only **mutation** makes new alleles.

Natural selection

Evolution is the gradual change in a group of organisms over time. Many theories have been suggested to explain how evolution may occur. The most popular is **natural selection** and was put forward by **Charles Darwin**.

Darwin made four important observations:
- Organisms often produce large numbers of offspring.
- Populations usually stay about the same size.
- Organisms are all slightly different; they show variation.
- Variation can be inherited.

Darwin used his observations to make these conclusions:
- More organisms are born than can survive.
- The best suited to the environment are the ones that survive and breed.
- They pass on their characteristics to their offspring.
- Over long periods of time the population changes.

If the two populations of the species become so different that they can no longer interbreed to produce fertile offspring, they have formed two new species.

Scientists think that natural selection has been happening for more than three billion years and has produced all species of living organisms from simple life forms.

6 Variation and natural selection

Variation

1 Write one letter in each box to indicate if the characteristic is controlled by genes only (**G**), environment only (**E**) or a combination of both (**B**).

a) A scar on the face

b) The masses of students in a class

c) Blood groups

d) Hair colour

2 a) What type of graph is usually used to plot data about characteristics that are controlled by genes and the environment? Explain your answer.

_____

b) What type of graph is usually used to plot data about characteristics that are controlled by one or two genes? Explain your answer.

_____

Natural selection

3 The photograph shows an Arctic hare.
They have smaller ears than other hares.

Here are some observations about the Arctic hare:

A Hares with small ears tend to give birth to small-eared hares.

B Hares with smaller ears lose less heat.

C Over thousands of years, the size of hares' ears has decreased.

D Hares produce many offspring but food is in short supply.

Write a letter in each box to match each observation with one of Darwin's conclusions.

a) More organisms are born than can survive.

b) Some organisms are better suited to the environment.

c) Many characteristics are inherited.

d) Over long periods of time the population changes.

4 A female tiger and a male lion can reproduce to produce a liger.
Explain why ligers are infertile.

_____

_____

Evidence for evolution and natural selection

Fossil evidence for evolution

There is evidence for **evolution** from looking at **fossils**. Fossils are the 'remains' of organisms from hundreds of thousands of years ago, which are found in rocks. Many of the organisms that formed fossils are not alive today. Their species have become **extinct**.

Fossils may be formed in various ways:
- from the hard parts of animals that do not decay easily
- from parts of organisms that have not decayed because one or more of the conditions needed for decay are absent
- when parts of the organisms are replaced by other materials as they decay
- as preserved traces of organisms, e.g. footprints, burrows and root pathways.

Scientists have used fossils to look at how organisms have gradually changed over long periods of time.

Although fossils have been very useful to scientists, there are problems:
- Many early forms of life were soft-bodied, which means that they have left few traces behind.
- What traces there were may have been destroyed by geological activity.

Evidence for natural selection

Trying to find evidence that evolution is the result of **natural selection** is difficult because evolution happens so slowly. Evidence can be seen in bacteria and insects because they reproduce quickly.

The development of antibiotic-resistant strains of bacteria can be explained using the theory of natural selection:
- When bacteria reproduce, mutations occur.
- Some mutated bacteria might be resistant to antibiotics.
- When antibiotics are used, these bacteria are not killed.
- They survive and reproduce, so a resistant strain develops.

The development of antibiotic resistance

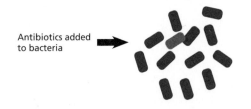

Antibiotics added to bacteria

Antibiotics kill off all susceptible bacteria

Only a single resistant bacterium survives

Bacterium multiplies to produce millions of resistant bacteria

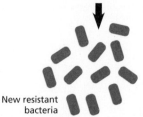

New resistant bacteria

How to avoid the development of antibiotic resistant bacteria is described on page 50.

Evidence for evolution and natural selection

Fossil evidence for evolution

1 What are fossils?

_____

_____

2 The diagram shows five fossils of a type of animal called an ammonite. The fossils are in different layers of rock in a cliff.

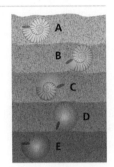

a) Which fossil, **A**, **B**, **C**, **D** or **E** is the oldest?

_____

b) Explain your answer to part **a)**.

_____

_____

3 **a)** Describe **one** way that fossils can be formed in rock.

_____

_____

b) Suggest why we do not have fossils of every stage of evolution.

_____

_____

Evidence for natural selection

4 Explain why it is possible to observe evolution occurring in bacteria.

_____

5 The drawing shows two types of peppered moth.

The peppered moth has been studied to provide evidence for natural selection.

Hundreds of years ago, most of the moths in an area were pale.

Occasionally a dark moth appeared.

Dark peppered moth   **Pale peppered moth**

a) Name the process that produces a dark moth.

_____

b) When the trees in the area were stained dark by pollution, scientists noticed that many more of the population of moths were dark.

Explain how natural selection could account for this change.

_____

_____

_____

Selective breeding and genetic engineering

Selective breeding

Humans have been using **selective breeding** for thousands of years. It is the process by which humans breed plants and animals for particular genetic characteristics.

Selective breeding involves several steps:
- Choose parents that best show the desired characteristic.
- Breed them together.
- From the offspring, again choose those with the desired characteristic and breed them.
- Continue over many generations.

The type of characteristic that could be selected for includes:
- disease resistance in food crops
- animals which produce more meat or milk
- domestic dogs with a gentle nature
- large or unusual flowers.

Selective breeding can lead to 'inbreeding'. This can happen in dogs, causing some breeds to be prone to disease or inherited defects.

What is genetic engineering?

Genetic engineering involves changing the characteristics of an organism by introducing a gene from another organism.

Genes from the chromosomes of humans and other organisms are 'cut out' and inserted into cells of other organisms. These genes will then be used to produce proteins.

These are the steps involved in genetic engineering:
- Enzymes are used to isolate and cut out the required gene.
- This gene is inserted into a vector, usually a bacterial plasmid or a virus.
- The vector is then used to insert the gene into the required cells.

If the genes are put into the cells of animals, plants or microorganisms at the egg or embryo stage then all cells in the organism will get the new gene.

Uses of genetic engineering

Microorganisms, plants and animals have all been changed by genetic engineering.

Plant crops have been genetically engineered to:
- be resistant to diseases, insects or herbicide attack
- produce bigger, better fruits
- contain increased levels of vitamins.

Crops that have had their genes modified by genetic engineering are called genetically modified (GM) crops.

Fungi or bacterial cells have been genetically engineered to produce useful substances such as human insulin to treat diabetes.

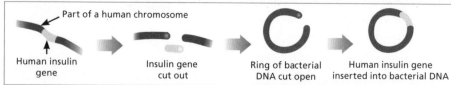

Part of a human chromosome

Human insulin gene — Insulin gene cut out — Ring of bacterial DNA cut open — Human insulin gene inserted into bacterial DNA

In humans, experiments are under way to use genetic modification to cure or prevent some inherited diseases.

Some people are concerned about genetic engineering. One view is that it is ethically wrong to change the genes of organisms.

Other concerns involve:
- the effects of GM organisms on populations of wildflowers and insects
- possible long-term effects on human health of eating GM crops.

Genetic engineering is a good example of a new technology that could be very useful but needs to have the risks investigated.

Selective breeding and genetic engineering

Selective breeding

1 Write down **two** uses of selective breeding for a farmer.

....

....

What is genetic engineering?

2 What is inserted into an organism when it is genetically engineered?

Put a ring around the answer.

| DNA | enzymes | insulin | protein |

3 Explain the role of each of the following in genetic engineering.

a) enzymes ....

....

b) plasmids ....

....

Uses of genetic engineering

4 Farmers use herbicides to kill weeds.

a) Explain how growing herbicide-resistant crop plants can allow a farmer to obtain a greater yield from his fields.

....

....

b) The pollen from herbicide-resistant crop plants might spread over a large area.

Why are some people concerned about this?

....

....

5 The diagram shows four stages in the production of human insulin by bacteria.

V

Insulin gene

Bacterial plasmid

W

X

Plasmid

Bacterium

Y

Human gene

Insulin gene

Put the stages **V**, **W**, **X** and **Y** in the correct order.

....

6 Classification

Classifying and naming

Organisms have been classified for thousands of years by putting them into groups.

One of the main systems that was used was developed by **Carl Linnaeus**:

- Linnaeus classified living things depending on their structure and characteristics.
- He put them into groups called kingdom, phylum, class, order, family, genus and species.
- He gave the organisms scientific names.

The scientific naming of organisms:

- is called the **binomial system**
- gives organisms a two-part name – the first part is the genus and the second is the species.

As microscopes improved, scientists learnt more about cells, and biochemical processes became better understood. Therefore, new models of classification were proposed.

Due to evidence such as genetic studies, there is now a **three-domain system**, which was developed by **Carl Woese**. In this system, organisms are first divided into:

- **Archaea** (primitive bacteria usually living in extreme environments)
- **Bacteria** (true bacteria)
- **Eukaryota** (which includes protists, fungi, plants and animals).

Example

The scientific name for lion is Panthera leo and for tiger it is Panthero tigris.

This shows that they are in the same genus but in different species.

The cheetah is called Acinonyx jubatus and so it is in a different genus and species.

Evolutionary trees

Evolutionary trees are a method used by scientists to show how they believe organisms are related. They use current classification data for living organisms and fossil data for extinct organisms.

The diagram shows the evolutionary tree for humans (*Homo sapiens*) and our closest relatives:

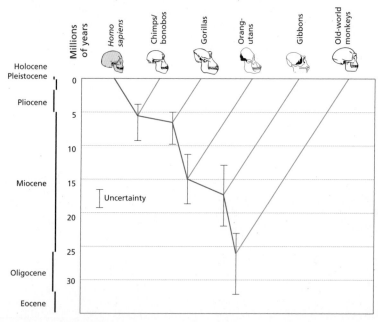

6 Classification

Classifying and naming

1 The table shows a classification of an animal called the serval cat.

Kingdom	Animalia
	Chordata
Class	Mammalia
	Carnviora
Family	Felidae
	Felis
Species	*serval*

a) Complete the table by writing the three different group names that are missing.

b) What is the scientific name for the serval cat?

_____

c) What is the name for the system that is used to give organisms scientific names?

_____

d) Name the scientist who invented this naming system.

_____

e) This table gives the common names of three different animals that are related to the serval cat.

Which of these cats is most closely related to the serval cat?

Explain your answer.

Common name	Scientific name
Jungle cat	*Felis chaus*
Pampas cat	*Leopardus colocola*
Fishing cat	*Prionailuris viverrinus*

_____

_____

f) Using the three-domain system, what is the name of the domain that includes the serval cat?

_____

g) What is the name of the scientist who developed the three-domain system?

_____

Evolutionary trees

2 Use the evolutionary tree on the facing page to answer these questions.

a) Which organisms on the tree are humans most closely related to?

_____

b) How many years ago did humans and orang-utans separate from a common ancestor?

_____

Interdependence

An **ecosystem** includes all the organisms living in a habitat and the non-living parts of their environment.

All the organisms of different species living in a habitat is called a **community**. A group of organisms of one species living in a habitat is called a **population**.

To survive and reproduce, organisms require a supply of materials from their habitat and from the other living organisms there. Trying to get these materials results in **competition**.

Plants and animals in a community compete for different things:

Plants compete for:	Animals compete for:
Light	Food
Space	Mates
Water	Territory
Mineral ions from the soil	

As well as competing with each other, species also rely on each other for food, shelter, pollination, seed dispersal, etc. This is called **interdependence**.

Due to interdependence, removing one species from a habitat can affect the whole community:

- In a stable community, all the species and environmental factors are in balance so that population sizes stay fairly constant.
- Tropical rainforests and ancient oak woodlands are examples of stable communities.

All ecosystems should be self-supporting, but they do need energy which is usually transferred in as light energy.

Tropical rainforest

Adaptations

Factors that can affect communities can be **abiotic** (non-living) or **biotic** (living).

Abiotic factors include:	Biotic factors include:
Light intensity	Availability of food
Temperature	New predators arriving
Moisture levels	New pathogens
Soil pH and mineral content	Competition with other species
Wind intensity and direction	
Carbon dioxide levels for plants	
Oxygen levels for aquatic animals	

Organisms have features (**adaptations**) that enable them to survive in the conditions in which they normally live.

- These adaptations may be structural, behavioural or functional.
- Some organisms live in environments that are very extreme, such as at high temperature, high pressure, or high salt concentration. These organisms are called **extremophiles**. Bacteria living in deep sea vents are extremophiles.

Cactii are adapted to live in extremely hot and dry environments

⑦ Interdependence and adaptations

Interdependence

① Draw lines to join each ecological term with its definition.

Ecological term	Definition
Community	An area where organisms live
Ecosystem	All the organisms living in an area
Habitat	All the members of a species living in an area
Population	The living and non-living parts of a habitat

② Tropical rainforests are very wet, warm habitats.

a) Complete these sentences about plants and animals by writing words in the gaps.

Orchids are plants that grow in tropical rainforests. It is very crowded in the forest, so plants

are competing for _____, which is needed for photosynthesis.

Bees in the forest compete with each other for territory, food and _____.

They receive some of their food from orchids, and in return pollinate the orchids.

This type of relationship is called _____.

b) What type of communities are tropical rainforests? Put a ring around the answer.

changing   stable   unstable   variable

Adaptations

③ Circle any abiotic factors in this list.

light   food   temperature   pathogens   predators

④ Cacti are plants that are adapted to live in hot deserts.
What name is given to organisms that live in extreme environments such as deserts?

⑤ Don Juan pond is a small lake in Antarctica.

It has been called the saltiest lake in the world and is twelve times more salty than normal sea water. Temperatures can be as low as -40°C in the lake.

Scientists claim they have found bacteria living in the water.

Give **two** reasons why it is so difficult for organisms to live in the lake.

Food chains

Feeding relationships in a community can be shown in **food chains**.

- All food chains begin with a **producer** which synthesises molecules. This is usually a green plant, which makes glucose by photosynthesis.
- Producers are eaten by **primary consumers**, which in turn may be eaten by **secondary consumers** and then **tertiary consumers**.
- Each of these feeding levels is called a **trophic level**.

Trophic levels can be represented by numbers, starting at level 1.

Consumers that eat other animals are called **predators**, and those who are eaten are called **prey**.

Top consumers are **apex predators** and are carnivores with no predators.

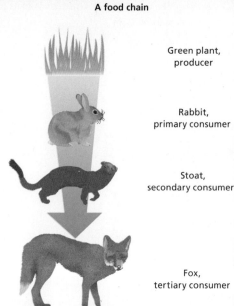

A food chain

Green plant, producer

Rabbit, primary consumer

Stoat, secondary consumer

Fox, tertiary consumer

Predator–prey graphs

In a stable community, the numbers of predators and prey often rise and fall in cycles. The numbers of predators and prey in a habitat will affect each other:

- The size of the two populations can be plotted on a graph.
- This is usually called a **predator–prey graph**.

In the graph:

- When the size of the prey population increases, there is more food for the predators.
- Therefore, the predator population increases.
- This means that more prey are eaten, so their numbers start to drop.
- This will be followed by a drop in the number of predators, as they have less food.
- Therefore, the prey numbers can increase again.

Predator–prey graph

In the predator–prey graph, both lines follow the same pattern but the peaks in predator numbers happen just after the peaks in prey numbers.

7 Feeding relationships

Food chains

1 A farmer is growing corn in a field. The corn can be fed on by locusts that are eaten by lizards. The lizards are eaten by snakes.

a) Draw a food chain for this field.

b) For this food chain write down:

  i) the source of energy ....

  ii) the apex predator ....

  iii) the primary consumer ....

c) How many trophic levels are there in this food chain?

....

d) How many predators are there in this food chain?

....

Predator–prey graphs

2 The graph shows a computer prediction for how the numbers of moose and wolves will change on a small island in the next 50 years.

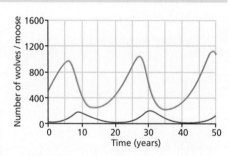

Wolf population ——
Moose population ——

a) What name is given to this type of graph?

....

b) Give **two** ways you can tell from the graph that the wolf is the predator in this relationship.

....

....

....

....

7 Cycles in nature

The carbon cycle

All materials in the living world need to be recycled so that the elements they contain can be used in future generations.

The **carbon cycle** describes how carbon is recycled in nature.

The carbon cycle involves **decomposers**:

- Decomposers are fungi and bacteria that feed on dead organisms or their waste products.
- The decomposers return carbon to the atmosphere as carbon dioxide from respiration. The carbon dioxide that has been released into the air can then be used by plants in photosynthesis.

Fossil fuels are being burnt at a faster rate than they are being formed, causing the carbon dioxide levels in the air to increase.

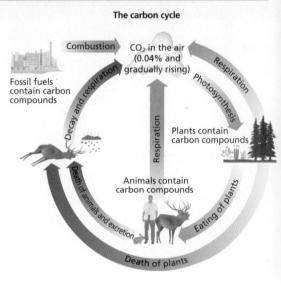

The carbon cycle

The water cycle

The **water cycle** describes how fresh water circulates between living organisms, rivers and the sea. It involves:

- **transpiration**, which allows water to be drawn up by plant roots and lost to the air from the leaves
- **evaporation**, which results in water on the ground being changed into water vapour in the air
- **condensation**, which converts the water vapour in the air into water droplets
- **precipitation**, which is the water falling back to the ground as rain or snow.

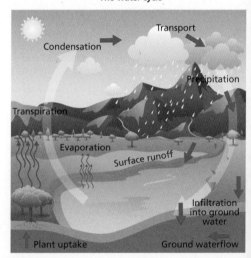

The water cycle

7 Cycles in nature

The carbon cycle

1. Name **two** types of organism that act as decomposers.

....

2. The diagram shows the carbon cycle.

   a) Name process **A** and process **B**.

   Process **A** = ...

   Process **B** = ...

   b) Add the letter **P** to the diagram to show where photosynthesis is happening.

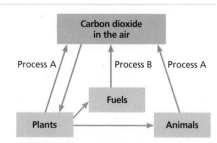

The water cycle

3. The diagram shows the water cycle.

   a) Complete the diagram by writing the names of the three missing processes in the boxes.

   b) Give the name of the process that returns water to the ground as rain or snow.

....

4. A plant loses water by transpiration.

   Explain why this transpiration is important to all the other plants in the ecosystem.

....

....

....

Biodiversity

Biodiversity is the variety of all the different species of organisms on earth, or in an ecosystem.

- A high biodiversity helps ecosystems to be stable, as species depend on each other for food and shelter.
- The future of humans on Earth relies on us maintaining a good level of biodiversity.

However, many human activities are reducing biodiversity and action is now being taken to try to stop this reduction.

> Due to the constant warm temperatures and plentiful rainfall, tropical rainforests have a very high biodiversity.

Pollution

Pollution kills plants and animals, which can reduce biodiversity.

Pollution can occur:

- in water, from sewage, fertiliser or toxic chemicals

- in air, from burning fossil fuels that release gases such as sulfur dioxide, which contribute to acid rain

- on land, from landfill and from toxic chemicals.

The human population is increasing rapidly and in many areas there is an increase in the standard of living. This means:

- more resources are being used
- more waste is produced.

The resources may be raw materials that may be extracted from the ground. Mining for minerals can damage ecosystems. There is also an increased demand for energy which can use resources such as fossil fuels.

The waste that is produced is sometimes burned, which can cause pollution. Alternatively it is dumped on land or at sea.

> Unless waste and chemical materials are properly handled, more pollution will be caused.

(7) Biodiversity and pollution

Biodiversity

1. Why does high biodiversity help an ecosystem to be stable?

2. A scientist collected samples of water of equal volume from two ponds, A and B.

   The diagram shows the animals that he found in the samples of water.

Pond A

Pond B

a) Which pond has the highest population of animals?
   How can you tell this?

b) Which pond has the highest biodiversity?
   How can you tell this?

c) Which pond is likely to be the most stable ecosystem?
   Explain your answer.

Pollution

3. The release of sulfur dioxide into the air has increased since 1880.

   a) Suggest why the level of sulfur dioxide release has increased.

   b) Name an environmental problem caused by sulfur dioxide release.

4. Give **one** reason why an increase in the size of human populations can lead to more pollution.

7 Global warming

Causes of global warming

Global warming is a gradual increase in the temperature of the Earth.

Many scientists think that it is being caused by changes in the levels of various gases in the atmosphere, due to pollution and deforestation:

- These gases are called **greenhouse gases**.
- They include carbon dioxide and methane.

This is how greenhouse gases warm the Earth:

- Radiation from the sun penetrates the Earth's atmosphere.
- This energy warms the Earth's surface.
- Some of this energy is re-radiated from the Earth's surface.
- This energy can now be absorbed by the greenhouse gases.
- This will warm up the Earth.

> Greenhouse gases keep the Earth at a temperature that allows life to exist. However, the levels of the gases are now increasing and this is raising temperatures to dangerous levels.

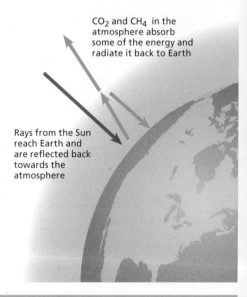

CO_2 and CH_4 in the atmosphere absorb some of the energy and radiate it back to Earth

Rays from the Sun reach Earth and are reflected back towards the atmosphere

Effects of global warming

There are a number of consequences of global warming including:

• more frequent **wildfires**	• longer periods of **drought** in some regions	• an increase in wind intensity and rainfall from **storms**.

These changes can affect animals and plants in different ways:

- loss of habitat when areas are flooded or burnt by wildfires
- changes in the distribution of species in areas where temperature or rainfall has changed
- changes to the migration patterns of animals
- changes in areas where pathogens can survive.

> Some tropical diseases such as malaria may be able to spread to other areas if the climate changes.

Causes of global warming

1. Put a ring around the **two** greenhouse gases in this list.

| carbon dioxide | methane | nitrogen | oxygen | sulfur dioxide |

2. Complete these sentences about global warming by writing words in the gaps.

Radiation from the .... penetrates the Earth's atmosphere.

This energy .... the Earth's surface but some is ....

This energy can be absorbed by .... gases and will further

.... the temperature of the Earth.

This process is called .... warming.

3. The graph shows the carbon dioxide levels in the air measured on a small, isolated island. It also shows changes in temperature.

a) Suggest why the carbon dioxide levels were measured on a small, isolated island.

....

....

....

....

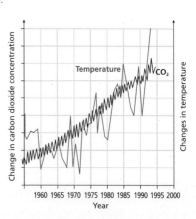

b) Why do the carbon dioxide levels go up and down each year?

....

....

....

Effects of global warming

4. The pathogen that causes malaria is carried by a mosquito that lives in tropical countries.
How could malaria be affected by global warming?

....

....

....

5. Climate changes might lead to certain plants being unable to live in some areas.
Explain why this could affect animal populations.

....

....

7 Overexploitation and conservation

Overexploitation

Humans put biodiversity at risk by causing pollution but also by taking too many resources out of the environment.

Overexploitation of the environment includes:
- building, quarrying, farming and dumping waste, which all reduce the amount of land available for animals and plants
- destroying **peat bogs** to produce garden compost, putting at risk the variety of different organisms that live there (the decay or burning of the peat will also release carbon dioxide into the atmosphere)
- cutting down trees and destruction of forests.

Destruction of forests is called **deforestation**.

In tropical areas, deforestation has:
- provided land for cattle and rice fields to produce more food
- provided land to grow crops to produce biofuels, based on ethanol.

Deforestation has:
- released more carbon dioxide into the atmosphere (because of burning and the activities of decomposers)
- reduced the removal of carbon dioxide from the atmosphere by photosynthesis, which means less carbon is 'locked up' in wood
- reduced the biodiversity of both plant species and the animals that live there.

Any increase in carbon dioxide levels will cause photosynthesis rates to increase. But, there is a limit to how well plants will be able to reduce the effect of global warming, especially if huge areas of rainforests are cut down.

Conservation

Scientists and governments have tried to reduce pollution and over-exploitation and so maintain biodiversity.

The steps used include:
- setting up breeding programmes for endangered species
- protecting rare habitats such as peat bogs, coral reefs, mangroves and heathland
- encouraging farmers to keep margins and hedgerows in fields
- reducing deforestation and carbon dioxide emissions
- recycling resources, rather than dumping waste in landfill.

In the UK, the sale of peat-based composts to gardeners was banned in 2024 but farmers can still buy it for several more years.

7 Overexploitation and conservation

Overexploitation

1 a) Explain how forests remove and store carbon from the air.

....

....

b) Give **two** reasons why deforestation is happening.

....

....

2 The graph shows the levels of two greenhouse gases, methane (CH_4) and carbon dioxide (CO_2), given out and taken in by three types of peat bog.

a) Use the graph to explain how natural peat bogs help to protect the Earth from global warming.

....

....

....

....

....

b) Peat bogs are often drained so that they dry out. Use the graph to explain why this is a bad idea.

....

....

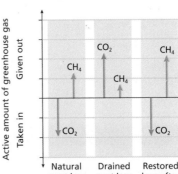

Conservation

3 Draw a line to link each conservation scheme to the threat to biodiversity that it may solve.

Scheme	Threat to biodiversity
Restoring hedgerows	Large areas of land used for landfill
Recycling rubbish	Populations of species decreasing to low numbers
Breeding programmes	Lack of nesting sites for birds

4 Conservation schemes are trying to restore peat bogs that have been drained.

Look at the graph in question 2 above. How effective are the schemes after 10 years?

....

....

....

The difference between atoms, ions and isotopes

Atoms

Substances are made of particles. An **atom** is the smallest particle that can exist on its own. In an element, all the atoms are all the same and there are just under 100 naturally occurring elements. All elements are listed on the **periodic table**.

Atoms are made of **subatomic particles**. Atoms of the same element have the same number of **protons** in the nucleus.

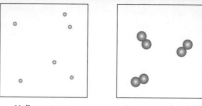

Helium atoms   **Oxygen molecules**

Every atom can be represented with a nuclear symbol. This gives the chemical symbol of the element as well as information about the subatomic particles it contains.

Parts of an atom

$$^{23}_{11}\text{Na}$$

Mass number
= number of protons + number of neutrons

Atomic number
= number of protons
= number of electrons in a neutral atom

Electron (negative electrical charge)

Proton (positive electrical charge)

Neutron (neutral)

Nucleus (protons and neutrons)

Ions

Ions are atoms with a charge. Metal elements:
- are on the left and towards the bottom of the periodic table
- form positive ions by losing **electrons**.

Non-metal elements:
- are on the right and the top of the periodic table
- become negative ions by gaining **electrons**.

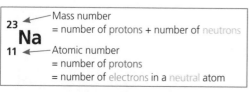

Neutral atom

Loss of electron(s)   Gain of electron(s)

Cation   Anion

Isotopes

Isotopes are different forms of an atom of an element. Isotopes of the same element have:
- the same **atomic number**
- a different **mass number**.

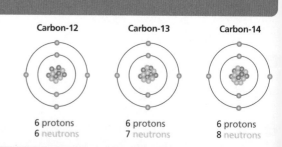

Carbon-12   **Carbon-13**   **Carbon-14**

6 protons
6 neutrons

6 protons
7 neutrons

6 protons
8 neutrons

Isotopes of the same element have the same number and arrangement of electrons. The electrons 'do' the chemistry and so they have the same chemical properties.

However, isotopes of the same element have a different mass number and so a different number of neutrons. This means that they have slightly different physical properties.

RETRIEVE
1

The difference between atoms, ions and isotopes

Atoms

1 What is an atom?

....

....

2 Why is hydrogen an element?

3 Approximately how many naturally occurring elements are there?

....

4 What are the names of the **three** subatomic particles?

....

Ions

5 What is an ion?

....

6 **a)** What charge do metal ions have?

....

b) What charge do non-metal ions have?

7 What sort of element is found on the left and towards the bottom of the periodic table?

....

Isotopes

8 What are isotopes?

....

....

9 **a)** How is the structure of isotopes of the same element similar?

....

....

b) How is the structure of isotopes of the same element different?

....

....

10 How many isotopes does carbon have?

....

The difference between compounds and mixtures

Compounds

Compounds are:
- more than one type of **atom** chemically joined
- formed from **elements** by **chemical reactions**.

Every compound has its own **formula**. A formula shows the symbols of the elements that it is made from and subscript numbers are used to show how many of each type of atom is present.

Compounds can only be separated into their elements by chemical reactions.

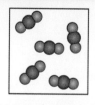

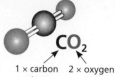

CO_2

1 × carbon atom   2 × oxygen atom

When the elements carbon and oxygen undergo a **chemical reaction (combustion)** the compound carbon dioxide is made.

Mixtures

Mixtures are made from more than one substance not chemically joined. In a mixture, the chemical properties of each substance are unchanged.

Some mixtures are designed to be useful products called **formulations**. They are usually made by mixing the components in carefully measured quantities so the product has the desired properties.

A mixture of two different gases

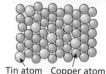

Tin atom   Copper atom

A metal alloy like bronze – an example of a formulation

Compounds melt and boil at specific temperatures but mixtures melt and boil over a range of temperatures. By measuring the melting point and boiling point of a substance, it is possible to classify a substance as a mixture.

Mixtures can be separated by a physical process as no new substances are made. Separation techniques include:

Chromatography e.g. Inks and dyes. Coloured inks from a pen: The number of colours can be seen and the same ink identified from more than one sample. 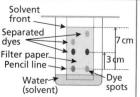 Solvent front, Separated dyes, Filter paper, Pencil line, Water (solvent), 7 cm, 3 cm, Dye spots	**Evaporation** e.g. Solutes from solvents in solutions. Sugar water: Small sugar crystals are formed and the solvent (water) is lost to the atmosphere. 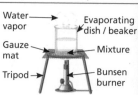 Water vapor, Evaporating dish / beaker, Gauze mat, Mixture, Tripod, Bunsen burner
Filtering e.g. Insoluble solid from a liquid. Sandy water: Sand is collected in the filter paper, and the water in the test tube. 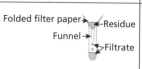 Folded filter paper, Residue, Funnel, Filtrate	**Crystallisation** e.g. Solutes from solvents in solutions. Sugar water: Large sugar crystals form in a saturated sugar solution. 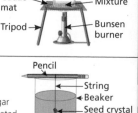 Pencil, String, Beaker, Seed crystal, Saturated solution
Distillation e.g. Solvent from solutes in solutions. Tap water: Pure water is collected in the beaker and solutes are left in the round-bottom flask. 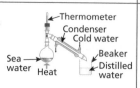 Thermometer, Condenser, Cold water, Beaker, Distilled water, Sea water, Heat	**Fractional distillation** e.g. Separating a liquid mixture where each liquid has a different boiling point. Crude oil: As the temperature increases, each fraction (liquid part) of the mixture evaporates and the vapour then condenses into the receiving flask. 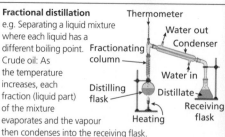 Thermometer, Water out, Condenser, Fractionating column, Water in, Distilling flask, Distillate, Receiving flask, Heating

The difference between compounds and mixtures

Compounds

1 In what type of reaction are compounds made?

2 **a)** How many atoms are in a molecule of carbon dioxide?

b) How many different elements are in a molecule of carbon dioxide?

3 In what type of reaction can compounds be separated back into their elements?

Mixtures

4 What is a mixture?

5 What is a formulation?

6 Which **two** elements are in bronze?

7 **a)** What sort of mixture can be separated by filtering?

b) Which separation technique would be used to make pure water from sea water?

c) How is distillation different to crystallisation? Give **one** similarity and **one** difference.

d) Which separation technique is used to separate crude oil into more useful mixtures?

The development of the model of the atom

Scientific models

Scientific **models** are simplified representations of what is really happening. Good scientific models can be used to understand **observations** and make **predictions**.

Over time, scientific models change or get replaced as new **data** (evidence) is collected from using new technology and through further investigations.

Early atomic models

Atoms were initially thought to be solid spheres that could not be divided.

As technology developed in the 1800s, the **electron** subatomic particle was discovered by J.J. Thomson. He modified the atomic model to form the plum pudding model of the atom, which has:
- a ball of positive charge
- negative electrons embedded in it.

Solid sphere model
(Dalton, 1803)

Plum pudding model
(Thomson, 1897)

Development of the model of the atom

Many scientists worked on trying to understand atoms better. In experiments by Geiger and Marsden, **alpha particles** (positive particles) were fired at thin sheets of gold foil. Most passed through, but some were deflected. They concluded that the atom was mainly empty space. Their data was used by Rutherford, who concluded that there was a small positive **nucleus** in the centre of the atom.

Niels Bohr (and others) used maths to show that electrons **orbit** the nucleus at specific distances. His theory was tested and agreed with real-life experimental **observations**.

The last subatomic particle to be found was the **neutron**, discovered by James Chadwick. Being uncharged, it was hard to detect in experiments.

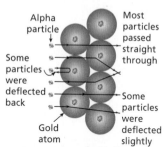

Alpha particle

Most particles passed straight through

Some particles were deflected back

Some particles were deflected slightly

Gold atom

The Geiger and Marsden experiment

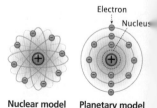

Electron
Nucleus

Nuclear model
(Rutherford, 1911)

Planetary model
(Bohr, 1913)

GCSE model of the atom

The GCSE model of an atom only works well for the first 20 elements.

Every atom has protons in the nucleus and electrons in the shells. Most atoms have neutrons in the nucleus. Hydrogen, ^1_1H, is the only atom that doesn't have any neutrons.

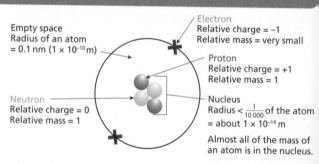

Empty space
Radius of an atom
= 0.1 nm (1×10^{-10} m)

Neutron
Relative charge = 0
Relative mass = 1

Electron
Relative charge = −1
Relative mass = very small

Proton
Relative charge = +1
Relative mass = 1

Nucleus
Radius < $\frac{1}{10\,000}$ of the atom
= about 1×10^{-14} m

Almost all of the mass of an atom is in the nucleus.

The development of the model of the atom

Scientific models

1 Why are scientific models useful?

_____

_____

2 Why do scientific models change?

_____

Early atomic models

3 Before the electron was discovered, how were atoms described?

_____

_____

4 What was J.J. Thomson's model of the atom called?

_____

Development of the model of the atom

5 What conclusion did Geiger and Marsden make from the results of their alpha scattering experiment?

_____

6 What conclusion did Rutherford make from the results of the alpha scattering experiment?

_____

_____

7 Which scientist discovered the neutron?

_____

GCSE model of the atom

8 Complete the sentences about the model of the atom by filling in the gaps.

Most of the mass of an atom is found in the _____.

0.1 nm or 1×10^{-10} m is the size of the _____ of an atom.

The subatomic particles found in the nucleus are _____ and

_____.

The charge of a proton is _____ and the charge of a neutron is

_____.

The mass of a neutron is _____.

Determining the subatomic particles in an atom

Nuclear symbols

Every atom can be represented with a nuclear symbol. This gives the chemical symbol of the **element** as well as information about the **subatomic particles** it contains.

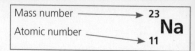

Mass number ⟶ 23

Atomic number ⟶ 11

$$^{23}_{11}\text{Na}$$

Calculating subatomic particles

For a neutral atom:
- the number of **protons** in the **nucleus** = **atomic (proton) number**
- the number of **electrons** in the shells = atomic number
- the number of **neutrons** in the nucleus = **mass number** – atomic number.

Each atom of the same element has the same number of protons and so the same atomic number. But some elements have **isotopes**, which have a different number of neutrons in the nucleus.

We can use the nuclear symbol to calculate the number of subatomic particles in an atom of an element.

Element	Nuclear symbol	Mass number	Atomic number	Number of protons	Number of electrons	Number of neutrons
Hydrogen	$^{1}_{1}\text{H}$	1	1	1	1	1 – 1 = 0
Helium	$^{4}_{2}\text{He}$	4	2	2	2	4 – 2 = 2
Lithium	$^{7}_{3}\text{Li}$	7	3	3	3	7 – 3 = 4
Beryllium	$^{9}_{4}\text{Be}$	9	4	4	4	9 – 4 = 5
Boron	$^{11}_{5}\text{B}$	11	5	5	5	11 – 5 = 6
Carbon	$^{12}_{6}\text{H}$	12	6	6	6	12 – 6 = 6
Nitrogen	$^{14}_{7}\text{N}$	14	7	7	7	14 – 7 = 7
Oxygen	$^{16}_{8}\text{O}$	16	8	8	8	16 – 8 = 8
Fluorine	$^{19}_{9}\text{F}$	19	9	9	9	19 – 9 = 10
Neon	$^{20}_{10}\text{Ne}$	20	10	10	10	20 – 10 = 10

Using each symbol on the periodic table, you:
- ✓ **can** work out the number of protons in the nucleus (atomic (proton) number)
- ✓ **can** work out the number of electrons in the shells (atomic number)
- ✗ **cannot** calculate the number of neutrons in the nucleus.

The periodic table does **not** display nuclear symbols. It shows the **relative atomic mass**, not the atomic mass, of each element:

relative atomic mass
atomic symbol
name
atomic (proton) number

Relative atomic mass is the mass of an 'average' atom of that element, considering the amount of each isotope present in a typical sample. This means that the relative atomic mass is not always a whole number (e.g. $^{35.5}_{17}\text{Cl}$).

It is not possible to have half a subatomic particle. So, the symbols in the periodic table cannot be used to calculate the number of neutrons in an atom of that element.

The atomic number is also called the proton number. It is the number of protons in an atom of an element. From this, you can infer that it is also the number of electrons in the neutral atom. In chemical reactions and physical changes, the atomic (proton) number of an atom or ion stays the same.

RETRIEVE 1 — Determining the subatomic particles in an atom

Nuclear symbols

1 a) What does the upper number of the nuclear symbol represent?

b) What does the lower number of the nuclear symbol represent?

Calculating subatomic particles

2 a) What is the name of the element with the nuclear symbol $^{23}_{11}Na$?

b) How many protons are in an atom of $^{23}_{11}Na$?

c) How many electrons are in an atom of $^{23}_{11}Na$?

d) How many neutrons are in an atom of $^{23}_{11}Na$?

3 What information can the atomic number give you about the number of subatomic particles?

4 How do you calculate the number of neutrons in an atom?

5 What information can a symbol in the periodic table give you about the subatomic particles in the atoms of a particular element?

6 Why is the symbol of an element in the periodic table **not** a nuclear symbol?

7 Define 'atomic (proton) number'.

The arrangement of electrons in the first 20 elements

Arrangement of electrons

Chemistry is the movement of **electrons** between atoms, ions or molecules. Electrons:

- fill each **energy level** (electron shell) in turn, starting with the one closest to the nucleus
- are kept in an **orbit** because of the **electrostatic attraction** between the negative electrons and the positive **nucleus**.

The electrons in the outer shell of an atom are responsible for its **chemical properties**.

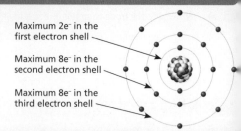

Maximum 2e⁻ in the first electron shell

Maximum 8e⁻ in the second electron shell

Maximum 8e⁻ in the third electron shell

Electronic structure of the first 20 elements

The **electronic structure** of these first elements can be shown in a diagram or by using digits (separated by commas, e.g. 2,1) to show how many electrons are in each energy level.

The group number is the same as the number of outer shell electrons of the atoms in that column of the periodic table. The period number is the same as the number of occupied electron shells of atoms in that row of the periodic table.

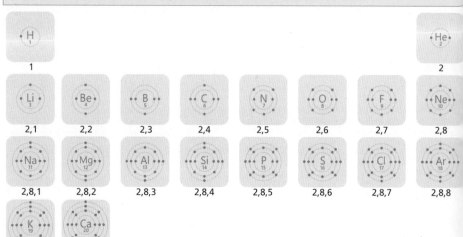

H 1							He 2
1							2
Li 3 2,1	Be 4 2,2	B 5 2,3	C 6 2,4	N 7 2,5	O 8 2,6	F 9 2,7	Ne 10 2,8
Na 11 2,8,1	Mg 12 2,8,2	Al 13 2,8,3	Si 14 2,8,4	P 15 2,8,5	S 16 2,8,6	Cl 17 2,8,7	Ar 18 2,8,8
K 19 2,8,8,1	Ca 20 2,8,8,2						

Electronic structure of ions

Ions are charged atoms with a full outer shell of electrons. **Metal** ions have fewer electrons than the metal atom and **non-metal** ions have more electrons than the non-metal atom.

The electronic structure of an ion is shown within a pair of square brackets with the charge written outside at the top right corner.

Sodium ion
2,8

Fluoride ion
2,8

$^{23}_{11}Na^+$
11 protons
10 electrons (11 – 1)
12 neutrons

Sodium is a metal. It forms positive ions when it loses an electron.

$^{19}_{9}F^-$
9 protons
10 electrons (9 + 1)
10 neutrons

Fluorine is a non-metal. It forms negative ions when it gains an electron.

RETRIEVE 1 — The arrangement of electrons in the first 20 elements

Arrangement of electrons

1 **a)** How many electrons complete the first energy level of an atom? ....

b) How many electrons complete the second energy level of an atom? ....

c) How many electrons complete the third energy level of an atom? ....

2 In what sequence do electrons fill energy levels in an atom?

....

....

Electronic structure of the first 20 elements

3 How many electrons do all Group 1 elements have in their outer shell? ....

4 How many electron shells are occupied in Period 3 elements? ....

5 Draw the electronic structure of a lithium atom.

6 Which element has atoms in the electronic structure 2,8,8? ....

Electronic structure of ions

7 What is an ion?

....

8 **a)** What is the charge on a sodium ion? ....

b) Draw the electronic structure of a sodium ion.

9 Explain why both sodium and fluoride ions have the same electronic structure.

....

....

....

....

The development of the periodic table

Classification

In science, objects with similar properties are **classified** (grouped) so that:
- they can be described easily without confusion
- connections can be seen between different objects
- **predictions** can be made about new objects that are found and fit into a particular group.

Scientists tried to organise **elements** into **groups** and classify them into tables before **subatomic particles** had been discovered. These tables were not very useful because:
- they were incomplete as not all elements had been discovered
- ordering was done using increasing atomic weight and some elements were placed in inappropriate groups.

Mendeleev's periodic table

In the 1800s, Dmitri Mendeleev overcame some of the problems of sorting the elements by:
- leaving gaps for elements still to be found
- broadly placing the elements in order of increasing atomic weight, but swapping some elements so that they were grouped by properties.

As technology improved, more elements were discovered and Mendeleev's predictions were shown to be correct. The discovery of subatomic particles and isotopes further supported the order of elements that Mendeleev developed.

The modern periodic table

There are some differences between Mendeleev's periodic table and the modern version:
- Information about the **atomic (proton) number** has been added for each element.
- **Atomic mass** is used rather than atomic weight.
- The four gaps left by Mendeleev have been filled

with the names, symbols and experimentally measured information about each element.
- An additional 50+ elements are included.
- An extra main column, Group 0 (**noble gases**), has been added.

Groups																		Periods
1	2												3	4	5	6	7	0	
												H						He	1
Li	Be												B	C	N	O	F	Ne	2
Na	Mg												Al	Si	P	S	Cl	Ar	3
K	Ca	Sc	Ti	V	Cr	Mn	Fe	Co	Ni	Cu	Zn	Ga	Ge	As	Se	Br	Kr	4
Rb	Sr	Y	Zr	Nb	Mo	Tc	Ru	Rh	Pd	Ag	Cd	In	Sn	Sb	Te	I	Xe	5
Cs	Ba	La	Hf	Ta	W	Re	Os	Ir	Pt	Au	Hg	Tl	Pb	Bi	Po	At	Rn	6
Fr	Ra	Ac	Rf	Db	Sg	Bh	Hs	Mt	Ds	Rg	Cn	Nh	Fl	Mc	Lv	Ts	Og	7

Metals
Non-metals

The elements in the modern periodic table are arranged by increasing atomic (proton) number, and in groups (columns) where elements have similar properties.

Most elements are **metals** and solid at room temperature (20°C). The position of an element in the periodic table can be used to predict its physical and chemical properties.

The properties of the elements repeat at regular intervals in the periodic table. Similar chemical properties are seen in elements of the same group because they have the same number of electrons in their outer electron shell.

(1) The development of the periodic table

Classification

1 What is meant by 'classification'?

....

....

2 Why do scientists classify objects?

....

....

....

....

3 Explain why early attempts to order the elements failed.

....

....

....

Mendeleev's periodic table

4 **a)** Describe how Mendeleev ordered the elements in the periodic table.

....

....

....

b) Why did Mendeleev leave gaps in his periodic table?

....

....

c) What new data justified Mendeleev's order of elements in the periodic table?

....

....

....

The modern periodic table

5 How are the elements arranged in the modern periodic table?

....

6 Which group of elements was added to the modern periodic table?

....

7 Complete this sentence.

Most elements can be classified as ....

Group 1 elements

Group 1 elements are **alkali metals**. They make up the first column of the periodic table. They:
- are metallic elements and so are **conductors**, **malleable**, **ductile** and **lustrous** (shiny)
- have **one electron** in the outer shell
- form **1⁺ ions** by losing the electron in their outer shell.

| Lithium atom 2,1 | Sodium atom 2,8,1 | Potassium atom 2,8,8,1 |

Properties of Group 1 elements

The **melting points** and **boiling points** of Group 1 elements decrease as you go down the group because there is a decrease in the force of attraction between the atoms.

Their **densities** increase as you go down the group.

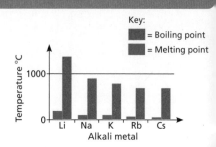

Key:
= Boiling point
= Melting point

Since all Group 1 elements have one electron in their outer shell, they have similar chemical properties. As you go down Group 1:
- atoms get larger
- there is less attraction between the outer-shell electrons and the positive nucleus
- atoms more easily form a 1⁺ ion
- the reactivity of the elements increases.

General equations can be used to summarise the common reactions of alkali metals:
- **alkali metal + oxygen → metal oxide**
- **alkali metal + water → metal salt + hydrogen**
- **alkali metal + chlorine → metal chloride**

Symbol	Reaction with oxygen	Reaction with chlorine	Reaction with water
Li	$4Li(s) + O_2(g) \rightarrow 2Li_2O(s)$ In a Bunsen burner, it burns with a red flame. Open to the air, it slowly tarnishes.	$2Li(s) + Cl_2(g) \rightarrow 2LiCl(s)$ Burns slowly with a red flame and a white solid is produced.	$2Li(s) + 2H_2O(l) \rightarrow 2LiOH(aq) + H_2(g)$ Floats and moves on the surface of the water as it effervesces (fizzes and bubbles). Gets smaller and seems to disappear.
Na	$4Na(s) + O_2(g) \rightarrow 2Na_2O(s)$ In a Bunsen burner, it burns with a yellow flame. Open to the air, it tarnishes quickly.	$2Na(s) + Cl_2(g) \rightarrow 2NaCl(s)$ Burns quickly with a bright yellow flame and a white solid is produced.	$2Na(s) + 2H_2O(l) \rightarrow 2NaOH(aq) + H_2(g)$ Floats and moves fast on the surface of the water as it effervesces. Melts into a sphere, gets smaller and seems to disappear.
K	$4K(s) + O_2(g) \rightarrow 2K_2O(s)$ In a Bunsen burner, it burns with a lilac flame. Open to the air, it tarnishes very quickly.	$2K(s) + Cl_2(g) \rightarrow 2KCl(s)$ Burns very brightly with a lilac flame and a white solid is produced.	$2K(s) + 2H_2O(l) \rightarrow 2KOH(aq) + H_2(g)$ Floats and moves very fast on the surface of the water as it effervesces. Ignites into a lilac flame, gets smaller and seems to disappear.

In chemical reactions, the alkali metal atoms lose one electron to become positive metal ions. This is an example of an oxidation reaction and can be shown in a diagram and a half-equation:

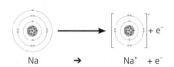

$$Na \rightarrow Na^+ + e^-$$

RETRIEVE

1 Group 1

Group 1 elements

1 Where are the Group 1 elements found in the periodic table?

2 What are the Group 1 elements also called?

3 Explain how the electron configuration is similar for all Group 1 metals.

4 What charge do Group 1 metal ions have?

Properties of Group 1 elements

5 **a)** What is the trend in the melting point as you go **down** Group 1?

b) Why does boiling point decrease as you go **down** Group 1?

c) What is the trend in density as you go **up** Group 1?

d) How does reactivity change as you go **down** Group 1?

6 What substance is made when rubidium reacts with oxygen?

7 What do you observe when sodium reacts with chlorine?

8 Describe what is meant by 'effervescence'.

① Group 7

Group 7 elements

Group 7 **elements** are in the seventh main column of the **periodic table**. Group 7 elements:

- are called the **halogens**
- are non-metal elements and so are **dull**, **poor conductors** and **brittle**
- have **seven electrons** in the outer shell
- form **1⁻ ions** by gaining an electron in the outer shell.

Fluorine atom 2,7   Chlorine atom 2,8,7

Properties of Group 7 elements

The Group 7 elements are made of **molecules** formed by a pair of halogen atoms, e.g:

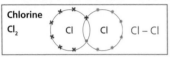

Chlorine Cl_2 Cl   Cl   Cl – Cl

The **melting point** and **boiling point** increase as you go down the group because there is a stronger force of attraction between the molecules. So, at room temperature (20°C), fluorine and chlorine are gases, bromine is the only non-metal liquid and iodine is a solid. As you go down the group, the halogens become darker in **colour**.

There is an increase in **relative formula mass** of the halogen molecule as you go down the group because the atoms get larger and heavier (they contain more subatomic particles).

All Group 7 elements have seven electrons in their outer shell. As you go down Group 7:

- the atoms get larger
- there is less attraction between the electrons in the outer shell and the positive nucleus
- atoms find it more difficult to form 1⁻ **halide** ions
- **reactivity** decreases.

Displacement reactions are when a more reactive halogen **displaces** a less reactive halogen from its **compound** (iodine (I_2) has no reaction).

Cl atom 2,8,7   Cl⁻ ion [2,8,8]

$+ e^- \rightarrow$

Chlorine, near the top of Group 7, is a highly reactive element that readily forms chloride (Cl⁻) ions

Halogen / Halide salt	Chlorine Cl_2 (aq)	Bromine Br_2 (aq)	Iodine I_2 (aq)
Sodium chloride NaCl (aq)	No reaction	No reaction	No reaction
Sodium bromide NaBr (aq)	*Observation*: Solution darkens. *Word equation*: chlorine + sodium bromide → sodium chloride + bromine. *Balanced symbol equation*: Cl_2 (aq) + 2NaBr (aq) → 2NaCl (aq) + Br_2 (aq). *Balanced ionic equation*: Cl_2 (aq) + 2Br⁻ (aq) → 2Cl⁻ (aq) + Br_2 (aq)	No reaction	No reaction
Sodium iodide NaI (aq)	*Observation*: Solution darkens. *Word equation*: chlorine + sodium iodide → sodium chloride + iodine. *Balanced symbol equation*: Cl_2 (aq) + 2NaI (aq) → 2NaCl (aq) + I_2 (aq). Balanced ionic equation: Cl_2 (aq) + 2I⁻ (aq) → 2Cl⁻ (aq) + I_2 (aq)	*Observation*: Solution darkens. *Word equation*: bromine + sodium iodide → sodium bromide + iodine. *Balanced symbol equation*: Br_2 (aq) + 2NaI (aq) → 2NaBr (aq) + I_2 (aq). *Balanced ionic equation*: Br_2 (aq) + 2I⁻ (aq) → 2Br⁻ (aq) + I_2 (aq)	No reaction

(1) Group 7

Group 7 elements

1 Where are the Group 7 elements found in the periodic table?

....

2 What are the Group 7 elements also called?

....

3 Explain how the electron configuration is similar for all Group 7 metals.

....

4 What charge do Group 7 non-metal ions have? ....

Properties of Group 7 elements

5 **a)** Describe the trend in the relative formula mass of the halogen molecule as you go down Group 7.

....

b) Why does boiling point increase as you go down Group 7?

....

....

c) What is the trend in the colour of the elements as you go down Group 7?

....

d) How does reactivity change as you go down Group 7?

....

6 What is a halide ion?

....

....

7 What is a halide salt?

....

8 Define the term 'displacement reaction'.

....

....

....

9 Write a word equation for the displacement reaction between chlorine water and a solution of sodium bromide.

....

....

1 Group 0

Group 0 elements

Group 0 **elements** make up the last column of the **periodic table**:

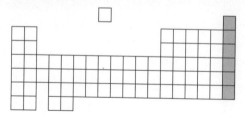

The Group 0 elements:
- are called the **noble gases**
- are non-metal elements
- have a full outer shell of **electrons**
- are all gases at room temperature.

Helium atom	Neon atom	Argon atom
2	2,8	2,8,8

Properties of Group 0 elements

There is a gradual change in **physical properties** as you go down the group.

Physical properties include:
- **Boiling point** – increases as you go down the group because there is an increased force of attraction between the atoms.
- **Relative atomic mass** – increases as you go down the group.

All the elements in Group 0 have a stable arrangement of electrons with a full outer shell:
- Helium has two electrons in the outer shell.
- Other noble gases have 8 electrons in the outer shell.

This **electronic stability** means that the Group 0 elements are **inert** (unreactive). So, noble gases:
- do not easily form **molecules** – they are found as gaseous single atoms (**monatomic**) at room temperature (20°C)
- do not usually react with other **non-metals**
- do not easily form ions
- do not usually react with **metals**.

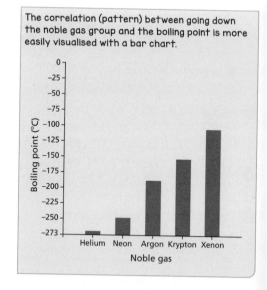

The correlation (pattern) between going down the noble gas group and the boiling point is more easily visualised with a bar chart.

1 Group 0

Group 0 elements

1 Where are the Group 0 elements found on the periodic table?

....

2 What are the Group 0 elements also known as?

....

3 Explain how the electron configuration is similar for all Group 0 elements.

....

....

4 What state are all Group 0 elements found in at room temperature (20°C)?

....

Properties of Group 0 elements

5 What is the trend in relative atomic mass as you go down Group 0?

....

6 Describe the relationship between relative atomic mass and boiling point.

....

....

7 a) How many electrons are in the outer shell of helium?
Tick the correct answer.

0 ☐   2 ☐   6 ☐   8 ☐

b) How many electrons are in the outer shell for all Group 0 elements except helium?
Tick the correct answer.

0 ☐   2 ☐   6 ☐   8 ☐

8 Why are Group 0 elements found as separate atoms?

....

9 Why do Group 0 elements rarely make compounds with:

a) another non-metal?

....

b) a metal?

....

② Covalent bonding

Covalent bonding

A **covalent bond** is formed when two atoms share a pair of electrons.

All chemical bonds are strong. This means that a lot of energy is needed to break them. When a covalent bond is broken, the atoms are separated.

Non-metals use **covalent bonds** to:
- complete their outer **shell** of **electrons**
- have a stable **noble gas electron configuration**.

A chlorine molecule (one covalent bond)

2 chlorine atoms ⟶

A chlorine molecule (made up of 2 chlorine atoms) ⟶

2 shared pair of electrons = covalent bond

Covalent bonding in elements and compounds

Other than the noble gases, most non-metal elements are found as molecules made up of two atoms. The table shows the **dot and cross diagrams** and **displayed formulae** for these **elements**.

	Chlorine, Cl_2	Hydrogen, H_2	Oxygen, O_2	Nitrogen, N_2
Dot and cross	Cl Cl	H H	O O	N N
Displayed formula	Cl – Cl	H – H	O = O (a double bond)	N ≡ N (a triple bond)

Many common substances make small molecules. The table shows the dot and cross diagrams and displayed formulae for these compounds.

Some elements, like diamond, make **giant structures** rather than molecules. Some compounds make giant structures like silicon dioxide, SiO_2.

	Water, H_2O	Hydrogen chloride, HCl	Methane, CH_4	Ammonia, NH_3
Dot and cross	H O H	H Cl	H C H	H N H
Displayed formula	H–O–H	H–Cl	H–C–H with H above and H below (H-C-H)	H–N–H with H below (H-N-H)

Polymers are very long molecules made from small repeating units called **monomers**. The **atoms** are held together by **covalent bonds**.

Polymers are made from **alkenes** (**monomers**). Ethene molecules can be shown with a chemical equation where n means lots of and the **repeating unit** is shown to represent the polymer:

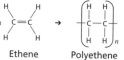

Ethene → Polyethene

Their structure is so long that we look for the small repeating unit and represent the whole structure by:
- writing the repeating structure in square brackets
- adding a lowercase 'n' to the outside bottom right of the bracket which means 'many of this unit'.

Representing structures

Scientific **models** can represent substances, but they have benefits and limitations:
- **Molecular formula** – only shows how many of which atom are present in the molecule.
- **Displayed formula** – shows every atom and every bond in the molecule but not shape or size of the molecule or the electrons.
- **Dot and cross diagram** – shows the electrons in relation to the atoms but does not show the shape or size of the molecule.
- **Ball and stick** – shows the number and type of atom present and suggests their size and position. But the atoms are too far apart from each other and the electrons are not visualised.
- **Diagrams** – 2D or 3D and show the number and arrangement of atoms, but no information about electrons. 3D computer models are useful as they rotate the structure.

② Covalent bonding

Covalent bonding

1 What sort of elements can form a covalent bond?

....

2 What is a covalent bond?

....

Covalent bonding in elements and compounds

3 Which non-metal group of elements on the periodic table do not undergo covalent bonding?

....

4 Draw the dot and cross diagram for a chlorine molecule.

5 How many covalent bonds are there in a nitrogen molecule?

....

6 Draw the displayed formula for an ammonia molecule.

7 Define the term 'polymer'.

....

....

Representing structures

8 What are the limitations of the ball and stick model to represent a covalent substance?

....

....

9 Which representation of covalent bonding shows the electrons? Tick the correct answer.

Dot and cross diagram ☐   Molecular formula ☐

Displayed formula ☐   Ball and stick model ☐

② Ionic bonding

Making metal ions and non-metal ions

Metal atoms lose all their outer shell **electrons** to get the **electronic structure of a noble gas**:

- Group 1 metals make 1+ ions, e.g. sodium, Na^+
- Group 2 metals make 2+ ions, e.g. magnesium, Mg^{2+}
- Group 3 metals make 3+ ions, e.g. aluminium, Al^{3+}

Non-metal atoms gain electrons to get a stable noble gas electronic configuration:

- Group 5 metals make 3- ions, e.g. nitride, N^{3-}
- Group 6 metals make 2- ions, e.g. oxide, O^{2-}
- Group 7 metals make 1- ions, e.g. chloride, Cl^-

Ionic equations are more challenging for non-metal atoms as they usually start as **molecules**.

Ionic equations are balanced in terms of charge and particle. So, for making a magnesium ion:

- Dot and cross diagram:

Mg atom 2,8,2 → Mg^{2+} ion $[2,8]^{2+}$ + $2e^-$

- Ionic equation: $Mg \rightarrow Mg^{2+} + 2e^-$

For making an oxygen ion:

- Dot and cross diagram:

- Ionic equation: $O_2 + 4e^- \rightarrow 2O^{2-}$

Forming an ionic bond

Ionic bonds are made when a metal reacts with a non-metal. The outer shell electrons on the metal atom are donated. The outer shell of the non-metal ion accepts the electrons. The resulting **electrostatic force** of attraction between the positive and negative ions is the ionic bond.

Often, only the outer shell electrons are shown in **dot and cross diagrams**:

Ionic compounds

Ionic compounds form a giant structure known as a **lattice**, where the electrostatic forces of attraction between oppositely charged ions act in all directions. So, 3D models can be useful at showing the arrangement of ions.

In an ionic compound, the charge is balanced. So, there may be more of one ion than another, e.g. magnesium chloride, $MgCl_2$:

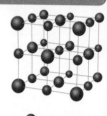

- ● Negatively charged chloride ions
- ● Positively charged sodium ions
- ● Cl^- ● Na^+

Cl^- ion $[2,8,8]^-$ Mg^{2+} ion $[2,8]^{2+}$ Cl^- ion $[2,8,8]^-$

To work out the ionic compound formula:
1. Write the formula of each ion next to each other, starting with the metal: Mg^{2+} Cl^{1-}
2. Circle the number of charges and cross them down:
   $Mg^{②} \times Cl^{①}$ becomes Mg_1Cl_2
3. Remove any 1s and cancel down: $MgCl_2$

The empirical formula is the smallest whole number ratio of atoms in a compound. The formula of any ionic compound is always the empirical formula of that compound.

② Ionic bonding

Making metal ions and non-metal ions

1 How do metal atoms become ions?

....

2 What is the charge on a Group 2 metal ion?

....

3 Write the formula for a sodium ion.

....

4 How do non-metal atoms become ions?

....

....

5 What is the charge on a Group 7 non-metal ion?

....

6 Write the formula for an oxide ion.

....

Forming an ionic bond

7 **a)** Define the term 'ionic bond'.

....

....

b) What types of elements can form an ionic bond?

....

c) How are electrons transferred to make an ionic bond?

....

Ionic compounds

8 What is the name of the structure formed by an ionic compound?

....

9 Give the formula for sodium oxide.

....

2 Metallic bonding

Metallic bonding in pure metals

Pure metals are substances that contain atoms of only one metallic **element**. The atoms are held together in a **giant structure** by **metallic bonds**.

The structure of a pure metal has:
- a giant structure of atoms in a regular pattern
- free moving **delocalised electrons**.

Metallic bonds are made when the outer shell electrons from the metal atoms leave the atom. So:
- the outer shell electrons become able to move freely in the structure
- a regular pattern of positive metal **ions** is left behind.

The sharing of the delocalised electrons between the metallic ions in the giant structure is a metallic bond.

The bonding in metals can be represented by 2D diagrams.

> All pure metals are solids at room temperature, except for mercury, Hg, which is a liquid.

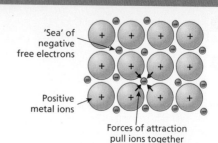

'Sea' of negative free electrons

Positive metal ions

Forces of attraction pull ions together

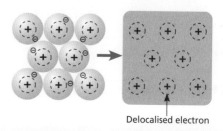

Delocalised electron

Properties of metals

The structure and bonding of a metal can be used to explain the general properties of metals:
- **Conductors** of electricity – delocalised electrons are free to move and carry the charge.
- **Ductile** (can be drawn into wires) and **malleable** (bends and shapes easily) – **planes** (layers) of atoms easily slide over each other.
- Conductors of heat – the delocalised electrons transfer the energy.
- High melting and boiling point – metallic bonds are strong and many bonds need to be broken to break the giant structure to melt or boil a metal; this requires a lot of energy.

Metallic bonding in alloys

Pure metals are usually quite soft and are not used in everyday applications.

The addition of another element with a different sized atom distorts the arrangement of atoms, stopping them from being able to slide as easily. These are **alloys**; they are harder and have more useful properties for everyday uses.

Alloys can be described as:
- a **mixture** of a metal and at least one other element
- a **formulation** (a mixture designed for a useful product).

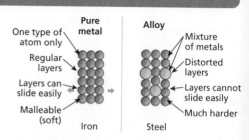

Pure metal
One type of atom only
Regular layers
Layers can slide easily
Malleable (soft)
Iron

Alloy
Mixture of metals
Distorted layers
Layers cannot slide easily
Much harder
Steel

> Metallic bonding can only occur in pure metals (metal elements) and alloys.

② Metallic bonding

Metallic bonding in pure metals

1 Define 'pure metals'.

....

....

2 **a)** What type of bonding is found in pure iron?

....

b) What sort of structure is found in pure iron?

....

c) How are the atoms arranged in a pure metal?

....

3 Explain what is meant by a 'metallic bond'.

....

....

Properties of metals

4 Draw lines to match the property on the left to the correct explanation on the right.

Property	Explanation
Conductor of heat	Many strong metallic bonds must be broken.
Malleable (bends and shapes easily)	Delocalised electrons are free to move and carry charge.
High melting and boiling points	The layers of atoms easily slide over each other.
Conductor of electricity	Delocalised electrons can transfer the energy.

Metallic bonding in alloys

5 What is an alloy?

....

....

6 Name an alloy made mainly from iron.

....

Giant structures and small molecules

Giant structures and their properties

Giant structures are 3D structures of **atoms** or **ions** held together by strong **bonds** so they have high melting points and a lot of energy is needed to break the bonds. There are three types:

Giant covalent structures	Giant ionic structures (lattices)	Giant metallic structures
Atoms are held by a shared pair of electrons.	Ions are held in position by the **electrostatic** force of attraction between the oppositely charged ions. Lattices are often soluble in water and can **conduct** electricity when the ions are free to move and carry the charge, so as a liquid or in **aqueous solution**.	Found only in **pure metal elements** or **alloys**. Metal ions are held in a regular arrangement by the attraction of **delocalised** electrons.
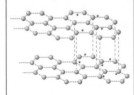	 Positively charged ion / Negatively charged ion	
Graphite is made of hexagonal rings that stack as layers	The structure of an ionic compound is an ionic lattice	Close-packed atoms

Small molecules and their properties

Molecules are small groups of **non-metal** atoms held together by covalent bonds. Many molecules are in a sample of a substance and there are weak forces of attraction between the molecules.

Strong covalent bond within the molecule

Weak forces of attraction between molecules

Simple molecules:

- are electrical **insulators** – no charged particles (ions or electrons) are free to move and carry a charge
- are soft and **brittle** – as the weak intermolecular forces of attraction are easily broken
- have low melting and boiling points – as only the weak forces of attraction between the molecules need to be overcome to melt or boil the substance. No strong covalent bonds are broken.

The melting point is always lower than the boiling point as a substance must melt before it can boil. As simple molecules get larger, the intermolecular forces of attraction also increase and so the melting and boiling points will be higher.

This relationship is easy to see in a **homologous series** (families of chemicals). Each successive **alkane** molecule has an extra $-CH_2-$ group compared to the last, and is heavier and larger. This increases intermolecular forces of attraction and so more energy is needed to overcome them, leading to higher melting and boiling points.

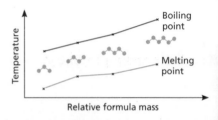

Group 0 elements have a stable outer shell of electrons. They have no bonds but do have weak forces of attraction between their atoms. As the atoms get larger, the weak forces of attraction increase. It is these weak forces of attraction that are overcome to melt and boil Group 0 elements. As these weak forces are very small, only a very small amount of energy is needed to overcome them and this is why these elements are found as gases at room temperature.

(2) Giant structures and small molecules

Giant structures and their properties

1 What holds the atoms together in a giant covalent structure?

....

2 What is the name given to giant ionic structures?

....

3 What sort of substances can make giant structures held by metallic bonds?

....

4 In which states can ionic compounds conduct electricity? Tick the correct answers.

Solid [ ]   Gas [ ]

Liquid [ ]   Aqueous solution [ ]

5 Explain why giant structures have high melting and boiling points.

....

....

....

Small molecules and their properties

6 What sort of atoms make molecules?

....

7 How does the size of a molecule affect the melting and boiling points of a substance?

....

....

8 Draw lines to match the property on the left to the correct explanation on the right.

Property

| Electrical insulator |

| Low melting and boiling points |

| Soft and brittle |

Explanation

| Only relatively weak intermolecular forces of attraction need to be overcome and no bonds are broken. |

| Weak intermolecular forces of attraction are easily broken. |

| No charged particles are free to move and carry a charge. |

② States of matter

States of matter

There are three states of matter, each with properties that can be explained by the particle model.

Each **atom**, or particle, of a substance has certain properties. These properties may not be the same as the **bulk properties**, which describe how a sample of the substance behaves.

The **particle model** allows us to explain observations about matter and make predictions about how matter will behave if conditions change. But there are limitations.

- There are no forces between the particles, whereas in real life there are bonds between the particles in giant structures, and intermolecular forces of attraction between molecules.
- All particles are the same shape (spheres). Although atoms tend to be spheres, molecules are a wide variety of shapes.
- All particles are solid, though some molecules are cages, e.g. buckminsterfullerene.

Solids
- Particles are touching and arranged in an order.
- Each particle vibrates around a fixed position.
- Fixed shape and volume.

Liquids
- Particles are still touching but they are in a random arrangement.
- As the particles can move past each other, liquids can flow.
- No definite shape but a fixed volume.

Gases
- Particles move randomly in all directions with different speeds.
- Gases spread out to fill the container (diffuse) and can be poured.
- No definite shape or volume.

Changing state

A substance can change state when energy is added or taken away. As no new substance is made, this is an example of a **physical change**.

Melting point and **boiling point** can predict the state at different temperatures.

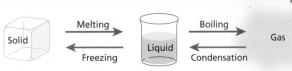

Energy transferred to the particles from the surroundings by heat

Solid — Melting → Liquid — Boiling → Gas

Solid ← Freezing — Liquid ← Condensation — Gas

Energy transferred from the particles to the surroundings by heat

Each substance has a different strength of attraction between its particles so each needs a different amount of energy to change state when melting and boiling. The stronger the forces of attraction between the particles, the more energy is needed for them to be overcome so the higher the melting and boiling points of the substance.

Substance	Melting point (°C)	Boiling point (°C)	State at room temperature
Bromine (Br$_2$)	-7	59	Liquid
Oxygen (O$_2$)	-219	-183	Gas
Water (H$_2$O)	0	100	Liquid
Sodium chloride (NaCl)	801	1465	Solid

State symbols

State symbols are shown after the formula of the substance in an equation, for example:
- Solid, e.g. ice = $H_2O(s)$
- Liquid, e.g. water = $H_2O(l)$
- Gas, e.g. steam = $H_2O(g)$
- Aqueous solution (a solution where water is a solvent), e.g. brine (sodium chloride solution) = NaCl(aq).

2 States of matter

States of matter

1 Complete the sentences about states of matter by filling in the missing words.

When a substance is a _____ or a _____ the

particles are able to move past each other.

When a substance is a _____ the particles are in an ordered pattern.

2 Explain why solids cannot be poured.

3 Which states of matter have a fixed volume?

4 Explain what the particle model is used for.

5 Give an example of a particle that is a cage and so not a solid sphere.

Changing state

6 Why is changing state described as a physical change?

7 Name the change of state when a substance changes from a solid to a liquid.

8 What happens when a substance condenses?

9 Ammonia has a melting point of -77°C and a boiling point of -33°C. What is the state of ammonia at room temperature?

State symbols

10 What is meant by '(aq)'?

11 What is the state symbol that you would use to show water at room temperature? _____

② The different structures of carbon

Carbon

Carbon is a **non-metal element** with the **electronic structure** 2,4. It is found in Group 4, Period 2 of the periodic table. It does not form any stable **ions** and so only makes **covalent bonds** in the elemental form with itself or with other non-metal **atoms** to form **compounds**.

> Carbon has many uses, but not every type of carbon is useful for every job.

Diamond, graphite and graphene

Diamond is a **giant covalent** structure. Each carbon atom is held in place by four strong covalent bonds.

- High melting and boiling points – many strong covalent bonds must be broken to pull apart the atoms.
- Electrical insulator (doesn't conduct electricity) – there are no charged particles (ions or **electrons**) that are free to move and carry the charge.
- Hard – the atoms are held in a ridged network.
- Used in drill bits – it is so hard and has such a high melting point that it can drill holes without being damaged itself.

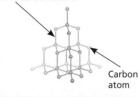

Covalent bond between two carbon atoms

Carbon atom

Graphite is a giant covalent structure. Each carbon atom is held in hexagonal layers by three strong covalent bonds. Each carbon atom donates one outer shell electron to form **delocalised electrons**, which move freely through the structure. There are weak forces of attraction between the layers of carbon atoms.

- High melting and boiling points – many strong covalent bonds must be broken to pull apart the atoms.
- Electrical conductor – delocalised electrons move freely and carry charge.
- Soft – the layers (**planes**) of atoms easily slide over each other.
- Used in pencils as the layers of carbon atoms easily slide off and leave a mark on the paper.

Covalent bond between two carbon atoms

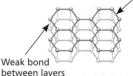

Weak bond between layers

Graphene is one layer of graphite. It is useful for making electronics and **composites**.

- High melting and boiling points – many strong covalent bonds must be broken to pull apart the atoms from the giant covalent structure.
- Electrical conductor – delocalised electrons are free to move and carry the charge.
- Flexible and strong – as the layers (planes) of atoms easily fold and bend while the strong covalent bonds do not break the atoms apart.
- Being developed into touch screens as highly flexible and an excellent electrical conductor.

Carbon atom

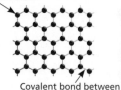

Covalent bond between carbon atoms

Fullerenes

Fullerenes are made only of carbon atoms and have a hollow shape. They are hexagonal rings of carbon atoms, but may also contain rings with five or seven carbon atoms. **Buckminsterfullerene** (C_{60}) was the first fullerene to be discovered and has a spherical shape.

Cylindrical fullerenes are called **carbon nanotubes**. They have very high length to diameter ratios and have properties that make them useful for **nanotechnology**, electronics and materials.

Buckminsterfullerene

Carbon atom

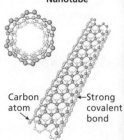

Strong covalent bond

Nanotube

Carbon atom

Strong covalent bond

2 The different structures of carbon

Carbon

1 How many electrons does carbon have in its outer shell?

2 **a)** What sort of bonding is found in a pure sample of carbon?

b) What sort of bonding is found in a compound containing carbon?

Diamond, graphite and graphene

3 **a)** How many strong bonds does each carbon atom have in diamond?

b) Why does diamond have very high melting and boiling points?

4 **a)** Where do the delocalised electrons come from in graphite?

b) How many strong bonds does each carbon atom have in graphite?

c) Why is graphite soft?

5 **a)** What is graphene?

b) Why can graphene conduct electricity?

Fullerenes

6 What is the formula of the first fullerene that was discovered?

7 What is **one** property of nanotubes?

Relative formula mass and percentage composition

Relative atomic mass of elements

The **relative atomic mass** of an **element** is the weighted average mass of an atom of that element, considering the natural abundance of **isotopes**. It has the symbol A_r and is found on the **periodic table** for every element.

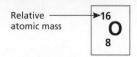
Relative atomic mass → 16
O
8

Relative formula mass

The **relative formula mass** of a **molecule** for a **non-metal** element is the sum of the atomic masses for each atom in the molecule, e.g. oxygen molecules (O_2) are made of two oxygen atoms **covalently** bonded together. So, the relative formula mass for a molecule of oxygen = $2 \times 16 = 32$.

The relative formula mass of a **simple covalent molecule** is the sum of the atomic masses for each atom in the molecule, e.g. water molecules (H_2O) are made of two hydrogen atoms covalently bonded to an oxygen atom.

Calculate the M_r of water, H_2O.

Write the formula	H_2O
Substitute the A_r values	$(2 \times 1) + 16$
Calculate the M_r	$2 + 16 = 18$

The empirical formula is the smallest whole number ratio of atoms in a compound.

The relative formula mass of an **ionic compound** is the sum of the atomic masses for the **empirical formula** of that compound, e.g.

Calculate the M_r of potassium carbonate, K_2CO_3.

Write the formula	K_2CO_3
Substitute the A_r values	$(39 \times 2) + 12 + (16 \times 3)$
Calculate the M_r	$78 + 12 + 48 = 138$

Sometimes the ratio of positive to negative ions is not 1:1. This is when brackets are used in formulae.

Calculate the relative formula mass of calcium nitrate, $Ca(NO_3)_2$.

Relative formula mass
= $40 + (14 \times 2) + (16 \times 6)$
= 164

Remember that everything inside a set of brackets is multiplied by the number outside the brackets, so $Ca(NO_3)_2$ contains 1 calcium, 2 nitrogen and 6 oxygen atoms.

Percentage composition

- Percentage is a mathematical comparison of how much you have of something compared to the whole, where the whole is 100.
- So, **percentage composition** =

$$\frac{\text{mass of atoms of a certain element in a compound}}{\text{relative formula mass of the compound}} \times 100$$

The formulae of all ionic compounds are empirical formulae as this is the unit that is repeated many times in the ionic lattice.

Consider the percentage composition of carbon in carbon dioxide:

What is the relative formula mass of carbon dioxide, CO_2?

Relative formula mass
= $(12 \times 1) + (16 \times 2)$
= 44

CO_2 contains 1 carbon atom with a relative atomic mass of 12, and 2 oxygen atoms with a relative atomic mass of 16.

So, the percentage composition of carbon in carbon dioxide = $\frac{12}{44} \times 100 = 27.3\%$

Relative formula mass and percentage composition

Relative atomic mass of elements

1 Where can you find the relative atomic masses of all the elements?

....

2 What is the relative atomic mass of carbon?

....

Relative formula mass

3 How do you calculate the relative formula mass for a molecule of an element?

....

....

4 **a)** What is the relative formula mass of hydrogen, H_2?

....

b) What is the relative formula mass of ozone, O_3?

....

5 How do you calculate the relative formula mass of a simple covalent molecule?

....

....

6 **a)** What is the relative formula mass of hydrogen peroxide, H_2O_2?

....

b) What is the relative formula mass of potassium oxide, K_2O?

....

7 How do you calculate the relative formula mass of an ionic compound?

....

....

Percentage composition

8 **a)** What is the percentage composition of oxygen in carbon dioxide, CO_2?

....

b) What is the percentage composition of hydrogen in water, H_2O?

....

c) What is the percentage composition of oxygen in water, H_2O?

....

③ Chemical and physical changes

Chemical and physical change

Chemical changes:
- are not easily reversible
- involve a new substance being made
- have no change in mass.

Physical changes:
- are reversible
- involve no new substance being made
- have no change in mass.

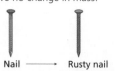

Nail ⟶ Rusty nail

Butter ⟶ Butter melts

Chemical reaction observations

- Gas formation
- Energy transfer (light, heat)
- Colour change
- Precipitate formation

Equations

Word equations give the names of the reactants and the products that are formed in a chemical reaction. Word equations should:
- include the names of all the chemicals involved in the reaction
- have an arrow (→) to show that the reactants go to the products

- have all the reactants on the left side of the arrow
- have all the products on the right side of the arrow.

Balanced equations show the formula of each substance involved in the chemical change. Atoms are not created or destroyed, just rearranged, so numbers are used to balance the equation.

Consider the reaction to make water from its elements:

hydrogen + oxygen → water

The reactants are two molecules of hydrogen and one molecule of oxygen.

The bonds in the reactants break to form two atoms of oxygen and four atoms of hydrogen.

The atoms rearrange and make new bonds to form the products which are two molecules of water.

$2H_2 + O_2 \rightarrow 2H_2O$
Two molecules of hydrogen ($2H_2$) react with one molecule of oxygen (O_2) to make two molecules of water ($2H_2O$).

Ionic equations show only the particles that are doing the chemistry. Ionic equations only have:
- the formulae of the **dissolved ions** directly involved in the chemical reaction
- the formulae of other **compounds** or **elements** that are not dissolved.

In the **neutralisation** reaction between hydrochloric acid and sodium hydroxide:
- hydrochloric acid + sodium hydroxide → sodium chloride + water
- Balanced symbol equation:
  $HCl(aq) + NaOH(aq) \rightarrow NaCl(aq) + H_2O(l)$
- Ions in solution:
  $H^+(aq) + Cl^-(aq) + Na^+(aq) + OH^-(aq) \rightarrow Na^+(aq) + Cl^-(aq) + H_2O(l)$
- Ionic equation (without the spectator ions):
  $H^+(aq) + OH^-(aq) \rightarrow H_2O(l)$

Half equations show either **reduction** or **oxidation** reactions. They are balanced in terms of particles and charge.

In a **displacement** reaction between zinc and copper sulfate:
- zinc + copper sulfate → zinc sulfate + copper
- Balanced symbol equation:
  $Zn(s) + CuSO_4(aq) \rightarrow ZnSO_4(aq) + Cu(s)$
- Ions in solution: $Zn(s) + Cu^{2+}(aq) + SO_4^{2+}(aq) \rightarrow Zn^{2+}(aq) + SO_4^{2+}(aq) + Cu(s)$
- Ionic equation (without the spectator ions):
  $Zn(s) + Cu^{2+}(aq) \rightarrow Zn^{2+}(aq) + Cu(s)$
- Half equation (for oxidation):
  $Zn \rightarrow Zn^{2+}(aq) + 2e^-$
- Half equation (for reduction): $Cu^{2+}(aq) + 2e^- \rightarrow Cu(s)$

Spectator ions are the ions that are unchanged in the solution where a reaction is happening.

If you add the half equation for oxidation with the half equation for reduction, you will get the balanced ionic equation for the redox reaction.

③ Chemical and physical changes

Chemical and physical change

1 Why is combustion (burning) classified as a chemical change?

....

2 Why is melting (changing state) classified as a physical change?

....

Equations

3 **a)** What are the substances at the start of a change called?

....

b) What are the substances at the end of a change called?

....

4 What does → mean in a word equation?

....

5 Why do symbol equations need to be balanced?

....

....

6 Balance the following chemical equation:

.... Mg + ...O_2 → .... MgO

7 What information do ionic equations give?

....

....

8 Define the term 'spectator ion'.

....

....

9 What do half equations show?

....

10 Write a half equation for the oxidation of copper atoms to form copper 2^+ ions.

....

Oxidation and reduction

Oxidation is an **exothermic** (gives out energy) chemical reaction. Oxidation is:
- addition of oxygen, e.g.
  magnesium + oxygen → magnesium oxide
- loss of **electrons**, e.g. $Mg \rightarrow Mg^{2+} + 2e^-$

Oxidation reactions include **combustion** and respiration.

Reduction is:
- loss of oxygen, e.g.
  iron oxide + carbon → iron + carbon dioxide
- gain of electrons, e.g. $2O^{2-} \rightarrow O_2 + 4e^-$

Reduction reactions include the extraction of metals from their ores.

> When one substance is oxidised, it does so by reducing the other. So, the whole reaction can be described as a REDOX reaction, because reduction and oxidation happen at the same time.

Neutralisation

Neutralisation is usually an exothermic chemical reaction. It is the reaction between an **acid** and a **base**, e.g.
- metal + acid → metal salt + hydrogen
- metal oxide + acid → metal salt + water
- metal hydroxide + acid → metal salt + water
- metal carbonate + acid → metal salt + water + carbon dioxide

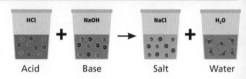

Acid   Base   Salt   Water

Neutralisation always produces a salt, e.g.
- hydrochloric acid makes chlorides
- sulfuric acid makes sulfates
- nitric acid makes nitrates.

Decomposition, precipitation and displacement

Decomposition reactions are **endothermic** (take in energy). There are two types of decomposition reaction:

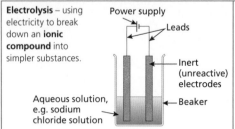

Electrolysis – using electricity to break down an **ionic compound** into simpler substances.

Power supply

Leads

Inert (unreactive) electrodes

Beaker

Aqueous solution, e.g. sodium chloride solution

Thermal decomposition – using heat energy to break down a substance into simpler substances, e.g. the thermal decomposition of copper carbonate to make copper oxide and carbon dioxide.

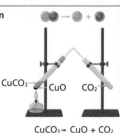

$CuCO_3$ CuO CO_2

$$CuCO_3 \rightarrow CuO + CO_2$$

Precipitation reactions are where two different solutions react to make a solid **insoluble precipitate**. This can be used as a method for making an insoluble salt. The precipitate can be removed from the reaction mixture by **filtering**.

Displacement reactions happen when a more reactive element takes the place of a less reactive element in its compound. Here are two examples:
- A more reactive halogen will take the place of a less reactive halogen in a compound.
- A more reactive metal will take the place of a less reactive metal in a compound.

> Some chemical reactions can be classified as more than one type of reaction, e.g. metal displacement is also a REDOX reaction as the more reactive metal is being oxidised and the less reactive metal is being reduced.

③ Types of chemical reaction

Oxidation and reduction

1 Why is combustion (burning) classified as an oxidation reaction?

....

2 What happens in terms of electrons when a substance is oxidised?

....

3 Why is extraction of iron from iron ore (iron oxide) classified as a reduction reaction?

....

4 What happens in terms of electrons when a substance is reduced?

....

Neutralisation

5 What are the **two** types of reactants in a neutralisation reaction?

....

6 What is always made in a neutralisation reaction?

....

Decomposition, precipitation and displacement

7 Name the **two** types of decomposition.

....

8 What sort of substance can undergo electrolysis?

....

....

9 Define the term 'precipitate'.

....

10 How can you separate the solid product in a precipitation reaction?

....

11 Describe what happens in a halogen displacement reaction.

....

....

12 Describe what happens in a metal displacement reaction.

....

....

Concentration and mass in reactions

Concentration of solutions

Many chemical reactions take place in **solutions**. Solutions are a **mixture** made of a **solute** (usually a solid) **dissolved** into a **solvent** (usually a liquid).

We can use the **particle model** to imagine the process of dissolving:
- The solute particles fit into the gaps in between the solvent particles.
- The **mass** of the solution = mass of solvent + mass of solute.
- The **volume** of the solution is usually the same volume as the solvent.

Concentration is a measure of how much solute is dissolved in 1 dm³ of a solvent.

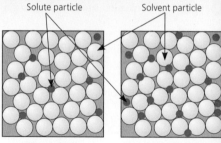

Solute particle   Solvent particle

Dilute solution
Not much solute in a given volume of solvent

Concentrated solution
Lots of solute in a given volume of solvent

Calculating concentration with mass

The concentration of a solution can be measured in mass per given volume of solution. Suitable **units** are grams per dm³, g/dm³ and gdm⁻³.

This **formula** can be used to work out the concentration of a solution:

$$\text{Concentration (g/dm}^3) = \frac{\text{mass (g)}}{\text{volume of solvent (dm}^3)}$$

2.00 dm³ of sodium hydroxide (NaOH) solution contains 20 g of sodium hydroxide.

Work out the concentration.
Concentration (g/dm³) =
20 g ÷ 2.00 dm³ =
10 g/dm³

Mass (g)

Conc. (g/dm³) | Volume (dm³)

Conservation of mass

In a chemical reaction, **mass** is conserved, as no atoms are gained or lost.

A **top pan balance** can measure the mass of the desirable product collected. Not every atom in the **reactants** will become a desirable product. This is due to one or more of these reasons:
- the reaction is **reversible**
- during each transfer, some product is lost
- the separation of the product from the reaction mixture leaves some product behind
- other reactions take place.

We can monitor the mass by using a **top pan balance**. We can see that the mass of the silver nitrate and sodium chloride at the start of the experiment is the same as the mass of the solution of sodium nitrate and the silver chloride precipitate made in the chemical reaction.

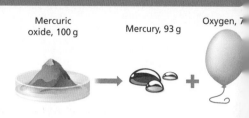

Mercuric oxide, 100 g   Mercury, 93 g   Oxygen, 7

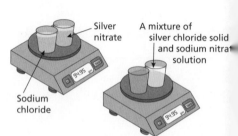

Silver nitrate   A mixture of silver chloride solid and sodium nitrate solution

Sodium chloride

Concentration and mass in reactions

Concentration of solutions

1 Draw lines to match the key term to the definition on the right.

| Solute | | The mixture |

| Solution | | The substance that dissolves |

| Solvent | | The substance that something dissolves in |

2 How would you describe a concentrated solution?

3 How would you describe a dilute solution?

4 Convert:

a) $0.5\,dm^3$ into cm^3

b) $1500\,cm^3$ into dm^3

Calculating concentration with mass

5 Write down the formula that connects concentration, mass and volume.

6 What is the concentration of a solution made from $10\,g$ of sodium hydroxide, NaOH, in $2.5\,dm^3$ solution?

Conservation of mass

7 What piece of measuring equipment can be used to measure the mass of a desired product?

8 Why doesn't every atom in the reactants become the desired product?

③ Moles

The mole

The **mole** is a measure of how many particles there are in a substance. The mole:
- is the chemical measure of the amount of substance
- is measured in **mol**
- has 6.02×10^{23} particles in it.

> In one mole of a substance there is the same number of particles per mole. As this number is always constant, it is called the **Avogadro Constant**.

For a mole of any element, the mass in grams is the same value as the relative atomic mass.

If you have one mole of a substance, the mass in grams is numerically equal to its relative formula mass. Consider the mass of one mole of sodium hydroxide, NaOH:

$$A_r \text{ sodium} + A_r \text{ hydrogen} + A_r \text{ oxygen}$$

$$= 23 + 1 + 16 = \textbf{40g}$$

The formula connecting amount of substance, mass and relative formula mass is:

$$\text{Number of moles of substance (mol)} = \frac{\text{mass of substance (g)}}{\text{mass of one mole (g/mol)}}$$

Example 1

Calculate the number of moles of carbon in 36g of the element.

$$= \frac{36g}{12 g/mol} = \textbf{3 moles} \quad \boxed{A_r \text{ carbon} = 12}$$

Example 2

Calculate the mass of 4 moles of sodium hydroxide. Rearrange the formula...

$$\text{Mass of substance (g)} = \frac{\text{number of moles}}{\text{of substance (mol)}} \times \frac{\text{mass of one}}{\text{mole (g/mol)}}$$

$$= 4\,mol \times 40\,g/mol = \textbf{160g}$$

Amount of substance in equations

Balanced symbol equations work on ratios. This can be considered in terms of moles, for example:
- Oxidation of magnesium:
    - magnesium + oxygen → magnesium oxide
    - $2Mg(s) + O_2(g) \rightarrow 2MgO(s)$

2 moles of Mg react with 1 mole of O_2 to produce 2 moles of MgO.

Work out how much MgO is made.
We know that the relative atomic mass of Mg is 24 and the relative formula mass of O_2 is 16×2 (= 32) and of MgO is 24 + 16 (= 40).

So if $2 \times 24g$ of Mg reacts with 32g of O_2 then $2 \times 40g$ of MgO is made.

$2Mg + O_2 \rightarrow 2MgO = 48g + 32g \rightarrow 80g$

If 4.8g of Mg is used then 8.0g of MgO is made.

Using moles to balance equations

In the oxidation of magnesium:

> 72 g of magnesium was reacted with 48 g of oxygen molecules to produce 120 g of magnesium oxide. Use the number of moles of reactants and products to write a balanced equation for the reaction.

amount of Mg $= \frac{72}{24} = 3\,mol$	Use the masses of the reactants to calculate the number of moles present.
amount of $O_2 = \frac{48}{32} = 1.5\,mol$	Divide the number of moles of each substance by the smallest number (1.5) to give the simplest whole number ratio.
amount of MgO $= \frac{120}{40} = 3\,mol$	This shows that 2 moles of magnesium react with 1 mole of oxygen molecules to produce 2 moles of magnesium oxide.

$$3Mg + 1.5O_2 \rightarrow 3MgO$$

$$2Mg + O_2 \rightarrow 2MgO$$

Often the expensive or most hazardous reactant is carefully measured to make sure that the limiting reactant can fully react and the maximum amount of product is made.

> In a chemical reaction, the reactant that is fully used up is called the **limiting reactant** and the reactant(s) that is not fully used up is described as being in excess.

Moles

The mole

1 Draw lines to match the key term on the left to the definition on the right.

Mole

Molar mass

Avogadro constant

Number of particles in a mole, 6.02×10^{23}

The mass of one mole of a substance, which is equal to the M_r value in grams

The chemical measure of the amount of substance

2 Write the formula that connects amount of substance, mass and relative atomic mass.

3 How many moles of oxygen atoms are in 32 g of oxygen molecules?

4 Write the formula that connects amount of substance, mass and relative formula mass.

Amount of substance in equations

5 How many moles of magnesium react with one mole of oxygen atoms in the oxidation of magnesium?

6 What mass of magnesium oxide is made when two moles of magnesium are completely oxidised?

Using moles to balance equations

7 What is a limiting reactant?

8 How do you ensure that a reactant is in excess?

4 The reactivity series

Metal reactions

Most **elements** are **metals**. When metal atoms react in a **chemical reaction** they:
- lose their outer shell **electrons**
- become positive **ions**
- are **oxidised**.

The reactivity of a metal is related to how easily it can become a positive ion.

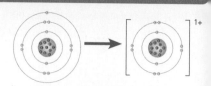

Sodium atom forming a sodium ion by losing an electron

The reactivity series

The **reactivity** of metals is a list of metals with:
- the most reactive metals at the top to the least reactive metals at the bottom
- the metals that form ions most easily at the top to the least at the bottom
- the most easily oxidised metals at the top.

The order of the metals in the reactivity series is based on the chemical reactions of metals with water and acids at room temperature (20°C).

Often the **non-metals** carbon and hydrogen are added to the reactivity series to predict reactions of metals:
- All metals above hydrogen in the reactivity series react with acids and those below do not.
- All metals above carbon in the reactivity series cannot be **extracted** from their **ores** by carbon.

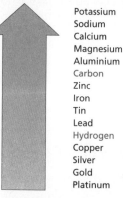

Most reactive

Potassium
Sodium
Calcium
Magnesium
Aluminium
Carbon
Zinc
Iron
Tin
Lead
Hydrogen
Copper
Silver
Gold
Platinum

Least reactive

Displacement reactions

Metal displacement reactions happen when a more reactive metal takes the place of a less reactive metal from its compounds.

magnesium + iron sulfate → magnesium sulfate + iron

Mg Fe SO₄ → Mg SO₄ Fe

The reactivity series can be used to predict the outcome of metal displacement reactions:
- Underline the names of the metals (pure metal and the metal in a compound) in the reactants.
- The metal that is highest in the reactivity series will be in the compound.
- If the most reactive metal is already in the compound in the **reactants**, there will be no reaction.

When copper is put in a solution of zinc sulfate, there is no reaction. This is because zinc is more reactive than copper and zinc is already in the compound.

If the most reactive metal is a pure metal in the reactants, there will be a reaction.

When zinc is put in a solution of copper sulfate, there is a displacement reaction. This is because zinc is more reactive than copper and zinc **displaces** (takes the place of) copper from its compound.

Metal atoms that are large lose their outer shell electrons more easily and are more reactive than small metal atoms. So, the reactivity of metals increases down a group.

The reactivity series

Metal reactions

1 What happens to the outer shell electrons when a metal atom reacts?

....

2 Write a half equation for the oxidation of a sodium atom.

....

The reactivity series

3 What is the reactivity series?

....

4 What chemical reactions are used to create the reactivity series?

....

....

5 Which **two** non-metals are often included in the reactivity series?

6 Give an example of a metal that will not react with water but will react with an acid.

....

Displacement reactions

7 Describe what happens in a metal displacement reaction.

....

....

8 What are the products when magnesium reacts with zinc sulfate?

....

9 Explain why there is no reaction when zinc is put into a solution of magnesium sulfate.

....

....

10 Write a word equation for the reaction of copper with magnesium sulfate.

....

11 Write a word equation for the reaction of magnesium with copper sulfate.

....

....

Metal as a resource

Metals are a **finite resource**. Metal ores and **unreactive** metals are mined and extracted, which causes **pollution** and habitat damage. But the negative impact of a mine can be reduced:
- Gases can be treated and **neutralised**.
- **Toxic** material can be specially stored to prevent pollution.
- Old mines can be made into nature reserves.

Unreactive metals like gold are found uncombined in nature. But most metals are found in metal **compounds** known as **minerals**, which need to undergo a **reduction** reaction to get the metal.

When it is economical to extract a metal from a mineral, the rock is described as an **ore**.

Reduction with carbon

Metals below carbon in the **reactivity series** can be reduced by carbon. The metal compound is mixed with carbon and heated. The reaction produces the metal and carbon dioxide (linked to **global warming** and **climate change**).

Haematite is the main iron ore containing iron oxide. Iron oxide undergoes carbon reduction in the **blast furnace**, where it reacts to make carbon monoxide, which then reduces the iron oxide to form impure iron. Limestone is also added to react with any acidic impurities to make a substance called **slag**. Slag can be used for breeze blocks and road building.

> Reduction with carbon gives an impure form of the metal, so electrolysis can purify the metal.

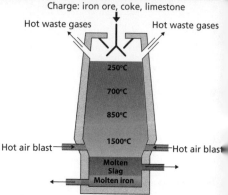

Charge: iron ore, coke, limestone

The extraction of iron is a displacement, and a redox reaction:

$$2\,Fe_2O_3 + 3C \rightarrow 4Fe + 3\,CO_2$$

Reduction (top), Oxidation (bottom)

Electrolysis

Electrolysis is the use of electricity to break down an ionic compound. It is used to extract metals when the metal is above carbon in the reactivity series and would react with carbon. The extraction process needs the metal-containing compound to be melted and an **electric current** passed through. Electrolysis produces a pure metal at the **cathode** and non-metals at the **anode**. Aluminium is extracted using electrolysis:
- Overall balanced symbol equation:
  $2Al_2O_3 \rightarrow 4Al + 3O_2$
- At the cathode (**reduction**): $Al^{3+} + 3e^- \rightarrow Al$
- At the anode (**oxidation**): $O^{2-} \rightarrow 2O_2 + 4e^-$

> Electrolysis is more expensive than metal extraction using carbon because it needs a lot of energy to melt the ionic compound.

The carbon anode needs to be continually replaced as the oxygen gas quickly reacts with the carbon of the anode: $C + O_2 \rightarrow CO_2$

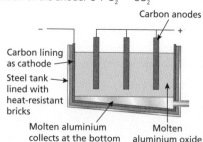

Bauxite is the main ore used to extract aluminium. The ore is processed to remove aluminium oxide, which is then mixed with **cryolite** to form the **electrolyte**. This has a lower melting point than a pure electrolyte so reduces the energy cost.

The extraction of metals

Metal as a resource

1 How are unreactive metals found in nature?

2 What is a mineral?

3 What is an ore?

4 Name the chemical reaction used to extract a metal from a metal compound.

Reduction with carbon

5 When can carbon be used to extract a metal from its compound?

6 Write a word equation for the extraction of iron from iron oxide using carbon.

7 Give the environmental problems linked to carbon dioxide.

Electrolysis

8 When is electrolysis used to extract metals?

9 Why is electrolysis an expensive method of metal extraction?

10 Write a word equation for the electrolysis of aluminium oxide.

11 Why does the carbon anode in aluminium electrolysis need to be continuously replaced?

4 REDOX reactions

Reduction

Reduction reactions happen when:
- oxygen is lost by a substance
- electrons are gained by a substance.

Examples of reduction reactions are:
- metal extraction with carbon or electrolysis
- displacement reactions

- in metal displacement reactions, the metal ion in the compound is reduced to form a metal atom
- in halogen displacement reactions, the halogen atom is reduced to form a halide ion.

Oxidation

Oxidation reactions happen when:
- oxygen is gained by a substance
- electrons are lost from a substance.

Examples of oxidation reactions are:
- **combustion**
- displacement reactions:
  - in metal displacement reactions, the metal atoms are oxidised to form metal ions in the compound
  - in halogen displacement reactions, the halide ions are oxidised to form halogen atoms.
- metals are oxidised to form metal ions in a salt when they react with acids.

Bubbles of gas rising to the surface

Magnesium strip

Dilute acid

Magnesium atoms are oxidised to magnesium ions in this reaction

REDOX

Reduction and oxidation do not happen on their own: when one substance is oxidised, the other substance is reduced. So, **REDOX** reactions are reactions where oxidation and reduction are happening at the same time.

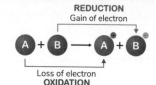

REDUCTION
Gain of electron

A + B ⟶ A + B

Loss of electron
OXIDATION

REDOX reactions include the following:
- Metal displacement reactions, e.g. the REDOX reaction between silver nitrate and copper:
  - The copper displaces the silver from silver nitrate – the **balanced symbol equation** is:
    $Cu(s) + 2AgNO_3(aq) \rightarrow Cu(NO_3)_2(aq) + 2Ag(s)$
  - As copper atoms lose electrons to become copper ions, copper being oxidised can be described using a **half equation**:
    $Cu \rightarrow Cu^{2+} + 2e^-$
  - As silver ions gain electrons to become silver atoms, silver being reduced can be described using a half equation: $Ag^+ + e^- \rightarrow Ag$

- Halogen displacement reactions, e.g. a reaction between potassium bromide and chlorine:
  - The chlorine displaces the bromide from potassium bromide – the balanced symbol equation is:
    $Cl_2(aq) + 2KBr(aq) \rightarrow 2KCl(aq) + Br_2(aq)$
  - As chlorine atoms gain electrons to become chloride ions, chlorine being reduced can be described using a half equation: $Cl_2 + 2e^- \rightarrow 2Cl^-$
  - As bromide ions lose electrons to become bromine atoms, which then form a diatomic molecule, bromide ions being reduced can be described using a half equation: $2Br^- \rightarrow Br_2 + 2e^-$

Oxidants are chemicals that supply oxygen and allow other substances to oxidise more easily. They can be hazardous and have a safety warning symbol. Oxidants undergo a REDOX reaction where they themselves are reduced.

④ REDOX reactions

Reduction

1 What happens, in terms of oxygen, in a reduction reaction?

....

2 What happens, in terms of electrons, in a reduction reaction?

....

3 Why is electrolysis of aluminium oxide classified as a reduction reaction?

....

4 Which substance is reduced in the reaction between magnesium metal and hydrochloric acid?

....

Oxidation

5 What happens, in terms of oxygen, in an oxidation reaction?

....

6 What happens, in terms of electrons, in an oxidation reaction?

....

7 Why is combustion of charcoal (impure carbon) classified as an oxidation reaction?

....

REDOX

8 What is a REDOX reaction?

....

....

9 Explain why a displacement reaction is classified as a REDOX reaction.

....

....

10 **a)** In the reaction between magnesium metal and copper sulfate, which substance is being oxidised?

....

b) In the reaction between magnesium metal and copper sulfate, which substance is being reduced?

....

c) Why is the reaction between magnesium metal and copper sulfate classified as a REDOX reaction?

....

4 Reactions of acids

Acids

Acids are substances that:
- have a pH <7
- release a $H^+(aq)$ **ion** in solution.

Three common acids you should know are:
- hydrochloric acid, HCl(aq)
- sulfuric acid, H_2SO_4(aq)
- nitric acid, HNO_3(aq).

Metals and acids

Metals above hydrogen in the **reactivity series** can react with acids to make a metal salt and hydrogen. The general equation for this reaction is:

metal + acid → metal salt + hydrogen

The name of the **metal salt** is found by looking at the name of the acid:

Acid	Salt
Hydrochloric acid	Metal chloride
Sulfuric acid	Metal sulfate
Nitric acid	Metal nitrate

In this reaction, metal atoms are **oxidised** to metal ions and **pH** increases. **Metal oxides** are **basic** and usually **insoluble**. A **neutralisation** reaction can take place between a metal oxide and an acid to produce metal salts and water. The general equation for this reaction is:

metal oxide + acid → metal salt + water

Metal hydroxides are basic and often they dissolve in water, so are **alkalis** too. A neutralisation reaction can happen between a metal hydroxide and an acid to make a metal salt and water. The general equation for this reaction is:

metal hydroxide + acid → metal salt + water

acid + base → salt + water

Metal carbonates are basic and usually insoluble. A **neutralisation** reaction can happen between a metal carbonate and an acid to produce a metal salt, water and carbon dioxide. The general equation for this reaction is:

metal carbonate + acid → metal salt + water + carbon dioxide

Salts

Salts are **ionic compounds** made when the hydrogen in an acid is swapped for a metal ion or the ammonium ion. The **solubility** of salts varies:
- Insoluble salts can be made from **precipitation** reactions and the salt can be separated by **filtering**.
- **Soluble** salts can be made from neutralisation reactions between an acid and an insoluble base (reactive metals, metal oxides, hydroxides or carbonates). **Excess** solid base is added to ensure the reaction is complete, then the filtrate is collected and the salt **crystalised**.

When salts are made, they are impure as they contain some solvent from the reaction mixture. To make a pure dry sample, the salt must be patted dry with absorbent paper or left in a drying oven.

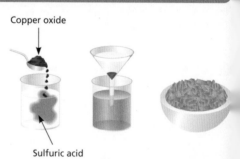

Copper oxide

Sulfuric acid

Add copper oxide to sulfuric acid → Filter to remove any unreacted copper oxide → Evaporate to leave behind blue crystals of the 'salt' copper sulfate

④ Reactions of acids

Acids

1 Which ion do acids release in aqueous solution?

....

2 Name the acid with the formula H_2SO_4.

....

Metals and acids

3 Which metals can react with acids?

....

4 Name the chemical reaction between metal oxides and acids.

....

5 What are the names of the products when calcium oxide reacts with sulfuric acid?

....

6 What are the names of the products when sodium hydroxide reacts with hydrochloric acid?

....

7 Write a word equation for the reaction between potassium hydroxide and sulfuric acid.

....

....

8 What gas is made when a metal carbonate reacts with an acid? Tick the correct answer.

Carbon monoxide ☐   Oxygen ☐

Carbon dioxide ☐   Hydrogen ☐

9 Write a word equation for the reaction between sodium carbonate and nitric acid.

....

....

Salts

10 Define the term 'salt'.

....

....

....

11 What chemical reaction can you use to make an insoluble salt?

....

(4) The pH scale

The pH scale

The **pH scale** is a measure of acidity or alkalinity of a **solution**. The scale starts at 0 and goes up to 14. The pH can classify a substance:

- **acids** have a pH <7
- **alkalis** have a pH >7
- **neutral solutions** have a pH = 7.

> The pH is a measure of the **concentration** of acid particles or hydrogen **ions** ($H^+(aq)$) in solution. The higher the **concentration** of these ions, the lower the pH value.

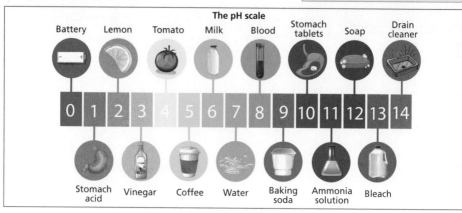

The pH scale

Battery | Lemon | Tomato | Milk | Blood | Stomach tablets | Soap | Drain cleaner

0 1 2 3 4 5 6 7 8 9 10 11 12 13 14

Stomach acid | Vinegar | Coffee | Water | Baking soda | Ammonia solution | Bleach

The pH scale is a **logarithmic scale**. This means that a solution with a pH of 1 has:

- × 10 more concentration $H^+(aq)$ when compared to a solution with a pH of 2
- × 100 more concentration $H^+(aq)$ when compared to a solution with a pH of 3

- × 1000 more concentration $H^+(aq)$ when compared to a solution with a pH of 4.

The pH of a solution can be changed by adding water and diluting the solution or by adding a different substance to neutralise some of the acid or alkali.

Measuring pH

pH can be measured with **universal indicator**:

- Add a few drops of universal indicator to the solution or dip the universal indicator paper into the solution.
- Compare the colour of the universal indicator to the colour chart and read off the pH value.

pH can be measured with a **pH probe**:

- Put the pH probe into the solution.
- Read off the pH value from the display.

Neutralisation

A **neutralisation** reaction happens when a base reacts with an acid. Alkalis are soluble **bases**. When an alkali reacts, the acid $H^+(aq)$ reacts with the alkali $OH^-(aq)$ to produce water.

This can be summarised in an **ionic equation**:

$$H^+(aq) + OH^-(aq) \rightarrow H_2O(l)$$

> A pH probe is a measuring instrument. If it is connected to a computer which can store the data, it would be called a datalogger.

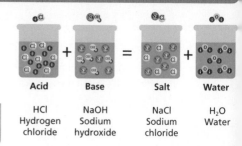

Acid | Base | | Salt | Water

HCl | NaOH | | NaCl | H_2O
Hydrogen chloride | Sodium hydroxide | | Sodium chloride | Water

The pH scale

1 Describe the pH scale.

....

....

2 Give the range that pH can be measured in.

....

3 How can you change the pH of a solution?

....

....

4 Draw lines to match the key term on the left with the pH value on the right.

Alkali	<7
Acid	>7
Neutral	=7

Measuring pH

5 Which substance can be used to measure the pH of a solution?

....

6 Which measuring instrument can be used to measure the pH of a solution?

....

Neutralisation

7 **a)** What sort of substance releases $H^+(aq)$?

....

b) What sort of substance releases $OH^-(aq)$?

....

c) What is made when $H^+(aq)$ and $OH^-(aq)$ react?

....

4 Strong and weak acids

Ionisation

Ionisation is the process of making an **ion**.

$$H{-}A \longrightarrow H^{\oplus} + A^{\ominus}$$

When **acids** are put into water, they release $H^+(aq)$ ions in **solution** and this is an example of ionisation. The more hydrogen ions there are in a solution, the lower the **pH** value.

This can be achieved by:
- having a more **concentrated** solution
- choosing an acid which is easier to ionise – the strength of the bond between the H and the rest of the substance varies with each acid (the easier it is to break the bond, the greater the ionisation).

Strong acids

Strong acids fully ionise in solution. This means that every acid formula dissociates into a $H^+(aq)$ ion and a negative counter ion. The **general equation** for this ionisation is: $HA \rightarrow H^+ + A^-$ (where HA is the formula of the acid, H^+ is the acid particle and A^- is the counter ion).

There are three strong acids that you need to know and show the ionisation for:
- Hydrochloric acid, $HCl \rightarrow H^+ + Cl^-$
- Nitric acid, $HNO_3 \rightarrow H^+ + NO_3^-$
- Sulfuric acid, $H_2SO_4 \rightarrow 2H^+ + SO_4^{2-}$

Strong and concentrated   **Strong and dilute**

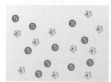

Weak acids

Weak acids like citric acid only partially ionise in aqueous solution. The general equation for the ionisation of a weak acid is: $HA \rightleftharpoons H^+ + A^-$

There are two weak acids that you need to know and show the ionisation for:
- Ethanoic acid, $CH_3COOH \rightarrow H^+ + CH_3COO^-$
- Carbonic acid, $H_2CO_3 \rightarrow H^+ + HCO_3^-$

Weak and concentrated   **Weak and dilute**

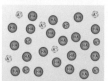
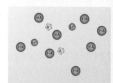

Concentration is a measure of how much solute is in a given volume of solution, whereas strength is a measure of how much a substance ionises. So, it is possible to have a concentrated solution of a weak acid or a dilute solution of a strong acid.

If you had two different samples of acid, one strong and one weak, the strong acid would have a lower pH value than the weak acid because there would be more hydrogen ions in solution.

Ethanoic acid

Key

(H) Hydrogen

(C) Carbon

(O) Oxygen

Carbonic acid

4 Strong and weak acids

Ionisation

1 What is ionisation?

....

2 How does the amount of $H^+(aq)$ affect pH?

....

....

Strong acids

3 **a)** What is a strong acid?

....

....

b) Give an example of a strong acid.

....

4 Write the general equation for the ionisation of a strong acid.

....

5 Write the balanced equation for the ionisation of hydrochloric acid.

....

Weak acids

6 **a)** What is a weak acid?

....

....

b) Give an example of a weak acid.

....

7 Write the general equation for the ionisation of a weak acid.

....

8 Write the balanced equation for the ionisation of ethanonic acid.

....

9 Write the balanced equation for the ionisation of methanoic acid (HCOOH).

....

....

Process of electrolysis

Electrolysis is:
- a **chemical change**, as something new is made
- an **endothermic change**, as energy is taken in by the system.

The process uses electricity to break down an **ionic substance** into simpler substances. The electrical **current** causes:
- positive metal **ions** or hydrogen ions to be attracted to the **cathode** (negative electrode)
- negative non-metal ions or hydroxide ions to be attracted to the **anode** (positive electrode).

Ions then become neutral atoms by gaining electrons (**reducing**) at the cathode and losing electrons (**oxidising**) at the anode.

For electrolysis to happen you need:
- **molten** or **aqueous solution** of an ionic compound
- a supply of **direct current** (e.g. from a battery)
- two **electrodes** (usually made of carbon).

You can monitor the electrolysis by having a **lamp** or **ammeter** in the circuit. When the circuit is complete and the current is flowing, the lamp will shine and the ammeter will have a reading.

Positive anode

Negative cathode

Electrolyte

Only when ions are free to move in electrolytes can electrolysis happen. This is because the ions need to carry the charge in the solution and complete the electrical circuit.

Uses of electrolysis

Electrolysis is used for:

Extracting metals from their compounds: metals like aluminium, that are higher than carbon in the **reactivity series**, will be extracted from their compounds using electrolysis.	Purifying metals: copper used in electrical wires must be very pure to reduce **resistance**. Impure copper is used to make the anode. The copper atoms become copper ions and enter the copper salt solution. Meanwhile, the copper ions in solution gain electrons at the cathode and become pure copper metal. The cathode is used to make the electrical wires.

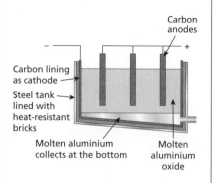

Carbon anodes

Carbon lining as cathode

Steel tank lined with heat-resistant bricks

Molten aluminium collects at the bottom

Molten aluminium oxide

Electrolysis of copper sulfate copper purification

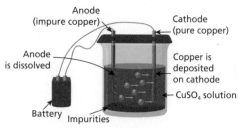

Anode (impure copper)

Cathode (pure copper)

Anode is dissolved

Copper is deposited on cathode

$CuSO_4$ solution

Battery Impurities

Anode: $Cu \rightarrow Cu^{2+} + 2e^-$
Cathode: $Cu^{2+} + 2e^- \rightarrow Cu$

④ Electrolysis

Process of electrolysis

1 a) Why is electrolysis an example of a chemical change?

....

b) Why is electrolysis an example of an endothermic change?

....

2 What sort of ions are attracted to the cathode?

....

3 What happens at the anode?

....

....

4 What sort of current is needed for electrolysis?
Tick the correct answer.

Alternating current (ac) ☐

Direct current (dc) ☐

5 What state must the electrolyte be in?

....

6 Which element are electrodes usually made from?

....

Uses of electrolysis

7 Name the metals that are extracted from their compounds using electrolysis.

....

....

8 In the extraction of aluminium, which electrode is the aluminium formed at?

....

9 Complete the sentence by filling in the gaps.

In the purification of copper, the mass of the anode .... and the mass of the

cathode ....

(4) Predicting the products of electrolysis

What happens at the anode and the cathode?

The **anode** is positive and attracts the negative **ions** in the **electrolyte**. Ions are **reduced** at the anode.

- In electrolysis of a melt, a **non-metal** is formed. The non-metal atoms will bond together to make **diatomic covalent molecules**.
- In electrolysis of a **solution**:
  - If the ionic compound doesn't contain a **halide**, oxygen is formed, e.g. $4OH^- \rightarrow O_2 + 2H_2O + 4e^-$
  - If the ionic compound does contain a halide, a **halogen** is formed, e.g. $2Cl^- \rightarrow Cl_2 + 2e^-$

The **cathode** is negative and attracts the positive ions in the electrolyte. Ions are **oxidised** at the anode.

- In electrolysis of a melt, a **pure metal** is formed.
- In electrolysis of a solution:
  - If the ionic compound contains a metal below hydrogen in the **reactivity series**, then the pure metal is formed, e.g. $Cu^{2+} + 2e^- \rightarrow Cu$
  - If the ionic compound contains a metal above hydrogen in the reactivity series then hydrogen is formed, e.g. $2H^+ + 2e^- \rightarrow H_2$

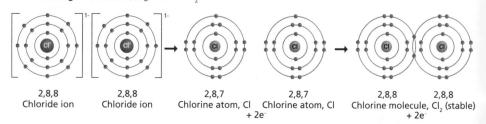

| 2,8,8 | 2,8,8 | 2,8,7 | 2,8,7 | 2,8,8 | 2,8,8 |
| Chloride ion | Chloride ion | Chlorine atom, Cl + 2e⁻ | Chlorine atom, Cl | Chlorine molecule, Cl₂ (stable) + 2e⁻ |

Each electrode attracts the oppositely charged ions from the electrolyte. Only one ion at each electrode can be discharged as an element. The reactivity of the elements determines what is produced.

Electrolysis of lead bromide

Molten lead bromide can be electrolysed in conditions available in a school lab as the melting point of lead bromide can be achieved using a Bunsen burner.

The overall electrolysis can be summarised by:
- word equation: lead bromide → lead + bromine
- symbol equation: $PbBr_2 \rightarrow Pb + Br_2$

The products of this reaction are:
- at the anode, bromide ions are oxidised to bromine atoms: $2Br^- \rightarrow Br_2 + 2e^-$
- at the cathode, lead ions are reduced to molten lead: $Pb^{2+} + 2e^- \rightarrow Pb$

Other products

In electrolysis of melts, the temperatures are so high that carbon electrodes can react with the non-metals that are formed, e.g. in aluminium extraction, the carbon anode burns away as it reacts with the oxygen formed to make carbon dioxide.

In electrolysis of solutions, the **spectator ions** form a third product. So, in the electrolysis of brine (NaCl(aq)), chlorine gas is formed at the anode, hydrogen gas at the cathode and a solution of sodium hydroxide is formed as well.

4 Predicting the products of electrolysis

What happens at the anode and the cathode?

1 Complete the sentences by filling in the gaps.

Non-metal substances are always formed at the ....

If a solution of a metal halide is electrolysed, a .... is formed at the anode.

If a solution of a metal sulfate is electrolysed, ... is formed at the anode.

2 Write a half equation for the production of bromine at the anode during the electrolysis of potassium bromide.

....

3 What type of substance is always formed at the cathode in the electrolysis of a melt?

....

4 What is formed at the cathode if a solution of sodium chloride is electrolysed?

....

5 What is formed at the cathode if a solution of copper sulfate is electrolysed?

....

6 Write a half equation for the production of copper at the cathode during the electrolysis of copper bromide.

....

Electrolysis of lead bromide

7 **a)** Give the formula of the element formed at the cathode during the electrolysis of molten lead bromide.

....

b) Give the formula of the element formed at the anode during the electrolysis of molten lead bromide.

....

Other products

8 Explain why the carbon anodes need to be replaced frequently in the extraction of aluminium with electrolysis.

....

....

....

⑤ Exothermic reactions

Energy

Energy can be transferred between the **chemical system** and the **surroundings**. The Law of Conservation of energy means:
- energy can neither be created nor destroyed
- the amount of energy in the Universe must remain constant
- the total energy in the system and the surroundings must be the same at the start and end of the change.

Exothermic reactions

Exothermic reactions release energy from the chemical system into the surroundings. So, the total **stored chemical energy** in the **reactants** is more than the total stored energy in the **products**.

The energy transferred to the surroundings causes a rise in **temperature**, which can be observed by using a **thermometer**.

The reaction mixture is put in an insulated vessel to reduce energy being lost to the surroundings. The thermometer measures the temperature change. The bigger the rise:
- the more energy has been released from the reactants as they become the products
- the more exothermic the reaction.

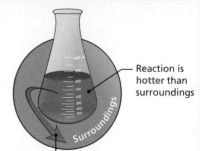

Reaction is hotter than surroundings

Surroundings

Reaction releases heat

Examples of exothermic reactions

Reaction profiles are diagrams that model how the energy changes from the start, during and at the end of the reaction.

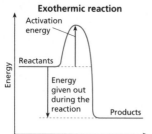

Exothermic reaction

Activation energy

Reactants

Energy

Energy given out during the reaction

Products

Progress of the reaction

In an exothermic reaction:
- reactants have more energy than the products
- the **activation energy** is the hill shape
- the difference between the reactants and the products is the energy released to the surroundings.

Most chemical reactions are exothermic reactions. They include:
- many **oxidation** reactions, including combustion
- **neutralisation**.

Everyday uses of exothermic reactions include:
- Hand warmers, where an exothermic change causes heat lost to the surroundings that can be used to warm a person.
- Self-heating cans that heat up food or drink: a button is pushed, which pierces the foil separator and allows liquid water to mix with solid quicklime. The exothermic reaction happens and a solution of calcium hydroxide is made. The chemical equation for the reaction is: $CaO(s) + H_2O(l) \rightarrow Ca(OH)_2(aq)$.

The heat given out is used to heat the food / drink in the container. (The chemicals from the reaction do not mix with the food / drink.)

The **activation energy** is the minimum energy required to start the reaction. This is usually a spark, flame or heat from friction to get an exothermic reaction to start. The activation energy of a reaction can be reduced by adding a catalyst, which gives an alternative pathway with a lower activation energy, so more collisions are successful and the rate of reaction increases.

⑤ Exothermic reactions

Energy

1 How does the total energy in the system and the surroundings compare at the start and end of a change?

....

Exothermic reactions

2 For an exothermic reaction, what happens to the energy from the chemical system?

....

....

3 In an exothermic reaction, how does the stored chemical energy of the reactants compare to the products?

....

....

4 What measuring instrument do you use to monitor temperature?

....

5 **a)** What is a reaction profile?

....

....

....

b) How do you use a reaction profile to calculate the energy released from an exothermic reaction?

....

....

c) How do the reactants compare to the products in an exothermic reaction's energy profile?

....

....

Examples of exothermic reactions

6 Give **two** everyday uses of exothermic reactions.

....

7 What is the formula of the liquid reactant used in the self-heating can?

....

8 What is the state of the product in the self-heating can? Tick the correct option.

Liquid ☐   Gas ☐   Aqueous solution ☐

5 Endothermic reactions

Endothermic reactions

Enothermic reactions do not release energy from the **chemical system**; they absorb energy from the surroundings. So, the total stored chemical energy in the **products** is more than the total stored energy in the **reactants**.

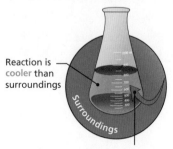

Reaction is cooler than surroundings

Surroundings

Reaction absorbs heat

The energy transferred from the surroundings causes a decrease in **temperature**, which can be observed by using a **thermometer**.

Endothermic reactions can be modelled in a reaction profile.

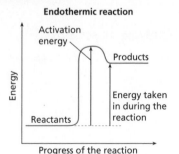

Endothermic reaction

In an endothermic reaction:
- products have more energy than the reactants
- the **activation energy** is from the reactant line to the top of the hill shape
- the difference between the reactants and the products is the energy taken in from the surroundings.

Examples of endothermic reactions

Endothermic reactions include **thermal decomposition** reactions where heat is used to break down a substance. For example, copper carbonate decomposes to copper oxide and carbon dioxide in an endothermic reaction.

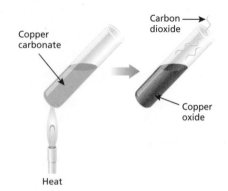

Copper carbonate

Carbon dioxide

Copper oxide

Heat

Everyday uses of endothermic reactions include:
- Sherbet is a formulation made from citric acid and sodium hydrogencarbonate, used in sweets. It mixes with water in the mouth and causes an endothermic neutralisation reaction forming a salt, water and carbon dioxide gas, which makes your mouth feel cool and tingly.
- When athletes injure themselves during sports they can use a cool pack, which contains two chemicals. When the outer bag is crushed, the inner bag of water bursts and mixes with the sodium ammonium nitrate. There is a physical change as the ammonium nitrate dissolves into the water to make a solution of ammonium nitrate. The equation for the change is:

$$NH_4NO_3(s) \xrightarrow{\text{H}_2\text{O}} NH_4NO_3(aq)$$

Energy is needed from the surroundings for **dissolving** to happen so this is an endothermic reaction.

⑤ Endothermic reactions

Endothermic reactions

① For an endothermic reaction, what happens to the energy from the surroundings?

_____

_____

② In an endothermic reaction, how does the stored chemical energy of the reactants compare to the products?

_____

_____

③ What happens to the temperature of the surroundings near an endothermic reaction?

_____

④ **a)** What is the x-axis on a reaction profile?

_____

b) What is the y-axis on a reaction profile?

_____

⑤ How do you use a reaction profile to calculate the energy released from an endothermic reaction?

_____

_____

Examples of endothermic reactions

⑥ Which chemical reaction is endothermic?
Tick the correct answer.

Neutralisation ☐

Thermal decomposition ☐

Oxidation ☐

⑦ Give **one** everyday use of endothermic reactions.

_____

⑧ Why is dissolving ammonium nitrate an endothermic reaction?

_____

_____

Modelling chemical reactions

A **scientific model** can be used to represent the changes happening in a **chemical reaction**:

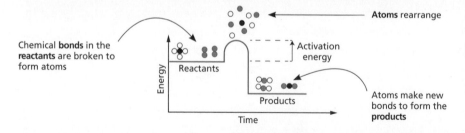

Chemical **bonds** in the **reactants** are broken to form atoms

Atoms rearrange

Activation energy

Reactants

Products

Time

Atoms make new bonds to form the **products**

Making and breaking bonds

Atoms are held together by chemical bonds.

Breaking bonds is an **endothermic** change. The stronger the bonds, the more energy is needed to break them.

Forming chemical bonds is an **exothermic** change. The stronger the bonds, the more energy is released when they are formed.

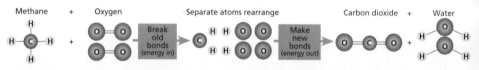

Methane + Oxygen → Separate atoms rearrange → Carbon dioxide + Water

Break old bonds (energy in)

Make new bonds (energy out)

Calculating the energy change of a reaction

The energy change of a reaction can be calculated from the **bond energy** data. Consider the **complete combustion** of methane:

$$CH_4 + 2O_2 \rightarrow CO_2 + 2H_2O$$

Energy needed to break the bonds in the reactants:

Type of bond	Number	Energy (kJ/mol)
C–H	4	4 × 412 = 1648
O=O	2	2 × 496 = 992
	Total	2640

Energy released when the bonds are formed in the products:

Type of bond	Number	Energy (kJ/mol)
C=O	2	2 × 805 = 1610
H–O	4	4 × 463 = 1852
	Total	3462

Overall energy change for a reaction = energy needed to break the reactant bonds − energy released when product bonds are made

2640 − 3462 = -822 kJ/mol

If the overall energy change is a negative number, then energy is released to the surroundings and the reaction is exothermic.

If the overall energy change is a positive number then the energy is taken in from the surroundings and the reaction is endothermic.

5 Bond energies

Modelling chemical reactions

1 a) In a chemical reaction, what happens to the bonds in the reactants?

b) In a chemical reaction, what happens to the bonds in the products?

Making and breaking bonds

2 Complete the sentences by filling in the gaps.

Breaking a bond is an .... change because energy is needed. Making a bond is

an .... change because energy is ....

3 How does the strength of the C=O bond compare to the strength of the C–H bond?

Calculating the energy change of a reaction

4 Give the unit that bond energy is measured in.

5 a) For an exothermic reaction, explain how the strength of all the reactant bonds compare to the strength of all the product bonds.

b) For an endothermic reaction, explain how the strength of all the reactant bonds compare to the strength of all the product bonds.

6 Write the formula to calculate the overall energy change of a reaction from bond energy data.

7 a) What is meant by 'a negative energy change for a reaction'?

b) What is meant by 'a positive energy change for a reaction'?

Measuring rate of reaction

The **rate** of a **chemical reaction** is:
- the speed of the chemical change
- defined by the change in the mass, concentration or volume of the **reactant** used or the mass, concentration or volume of the **product** formed in a given time.

> The mean (average) rate of reaction
> $$= \frac{\text{quantity of reactant used}}{\text{time taken}}$$
> $$= \frac{\text{quantity of product formed}}{\text{time taken}}$$

Some chemical reactions appear to have a **mass** change and can be monitored by using a **top pan balance**. The mass appears to change because:
- a reactant is a gas found in the air, causing the mass to appear to increase
- a product is a gas lost to the air, causing the mass to appear to decrease.

So, rate of reaction would be measured in g/s.

Some chemical reactions produce a gas, which can be collected either by **displacement** of water or in a **gas syringe**. The **volume** is measured, so rate of reaction would be measured in cm³/s. It is also possible to monitor the change in amount of one of the substances. In this case, the unit for rate of reaction would be mol/s.

Some chemical reactions make an **insoluble precipitate**, which makes the reaction mixture **turbid** (cloudy). You can observe a chemical reaction and time how long it takes until you cannot see through the solution. This method is known as the disappearing cross.

Displacement of water

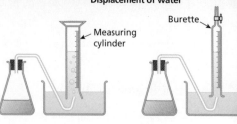

Measuring cylinder

Burette

Using a gas syringe

Gas syringe

The disappearing cross

Timer

Add dilute hydrochloric acid

Sodium thiosulfate

Flask

Paper with cross drawn on it

Calculating rate of reaction

Numerical data can be collected from experiments and plotted on a graph:
- The **independent variable** is time and is plotted on the x-axis.
- The **dependent variable** is either volume or mass and is plotted on the y-axis.
- The **gradient** gives the rate of reaction at that point. The higher the gradient / steeper the curve, the faster the rate of reaction.

The graph shows that reaction A is faster than reaction B.

To calculate the gradient of a point on a curve, you first need to draw a tangent to the curve. Then calculate the gradient of the tangent.

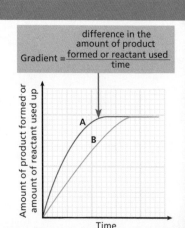

$$\text{Gradient} = \frac{\text{difference in the amount of product formed or reactant used}}{\text{time}}$$

Amount of product formed or amount of reactant used up

A

B

Time

Measuring and calculating rate of reaction

Measuring rate of reaction

1 Explain what is meant by 'rate of reaction'.

2 What measuring instrument do you use to measure mass?

3 **a)** Why might the mass of a chemical reaction appear to increase?

b) Why might the mass of a chemical reaction appear to decrease?

4 What is meant by 'turbid'?

5 How could you monitor the rate of reaction between sodium thiosulfate and hydrochloric acid, which produces a cloudy liquid?

6 What are the **two** units that rate of reaction can be measured in?

Calculating rate of reaction

7 When using a graph to determine rate of reaction, what will the independent variable be?

8 What is the relationship between the gradient of a rate of reaction graph and the rate of reaction?

9 What does it mean when the rate of reaction trend line is horizontal (gradient = 0)?

Collision theory

Collision theory

Collison theory is a **scientific model** that can be used to explain:
- how chemical reactions happen
- the effect of changing conditions on a chemical reaction.

For a chemical reaction to happen, a collision must happen between the **reactants**, in the correct orientation and with enough energy to be equal to, or higher than, the **activation energy**. So, most collisions do not lead to a reaction.

Increasing rate of reaction

Increasing **temperature** increases the rate of reaction. This is because the particles:
- move faster, increasing the number of collisions per unit time
- have more energy so each collision is more likely to be equal to or higher than the activation energy.

Cold reaction   Hot reaction

Increasing **surface area** of a solid reactant increases the rate of reaction. This is because:
- more of the solid particles are available for collision at any one time
- the number of collisions in a given time increases, but doesn't affect the percentage of successful collisions.

Large particle   Small particles
(smaller surface area) (larger surface area)

Increasing **concentration** of a reactant in solution increases the rate of reaction. This is because:
- more reactant particles are available for collision at any one time
- the number of collisions in a given time increases, but doesn't affect the percentage of successful collisions
- as a reaction happens, the concentration of the reactants reduces and the concentration of products increases.

Low   High
concentration concentration

🔵 Reacting particle of substance **A**
🔵 Reacting particle of substance **B**

Adding a **catalyst** increases the rate of reaction as:
- it provides an alternative pathway for the reaction to happen
- it reduces the activation energy
- although there is the same number of collisions in a given time, more are successful.

Energy

Reactants

Activation energy without catalyst

Activation energy with catalyst

Products

Progress of reaction

Increasing **pressure** of a gaseous reactant increases the rate of reaction. This is because:
- pressure is like the concentration of a gas, so if the pressure of a gaseous reactant increases, there are the same number of particles available but they are in a smaller volume so more likely to collide.

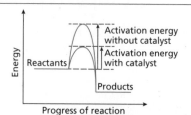

Increase in pressure increases rate of reaction ➡

Any reaction has the fastest rate of reaction at the start, slows down and stops at the end.

Different reactions need different catalysts. A catalyst in a **biological system** is an **enzyme**.

Collision theory

Collision theory

1 What is collision theory?

....

....

2 What does collision theory state is needed for a chemical reaction to happen?

....

....

....

Increasing rate of reaction

3 Complete the sentence by filling in the gaps.

When the temperature is increased, the rate of reaction .... and the

overall number of collisions in a given time ....

4 Explain why there are more successful collisions in a given time if surface area of a solid reactant is increased.

....

....

....

....

....

5 What happens to the rate of reaction if you increase the concentration of a reactant in solution?

....

6 What is the name given to a biological catalyst?

....

7 If you add a catalyst to a reaction, what happens to the overall number of collisions in a given time?

....

8 Which diagram shows increased temperature during a reaction?

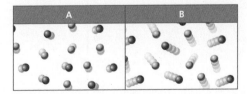

6 Reversible reactions

Reversible reactions

Some chemical reactions are described as reversible where the:
- reactants make the products
- products make the reactants.

Reversible reactions are summarised in chemical equations, but instead of having a normal arrow, this arrow is used: $\rightleftharpoons$.

Forward reaction

Backward reaction

The direction of the reversible reaction can be changed by changing the conditions. Conditions include **temperature**, **amount of substance** and **pressure**.

The energy change of a forward or backward reversible reaction is the same, but the energy is either taken in from the surroundings (endothermic) or released to the surroundings (exothermic). For example, hydrated copper(II) sulfate can thermally decompose and takes in 77 kJ/mol of energy; then when anhydrous copper(II) sulfate becomes hydrated, it releases 77 kJ/mol of energy.

hydrated copper sulfate	endothermic reaction $\rightleftharpoons$ exothermic reaction	anhydrous copper sulfate (white) + water

Ammonium chloride is a white powder that can thermally decompose to form ammonia and hydrogen chloride gas. As the gases cool, they easily undergo a neutralisation reaction to reform the ammonium chloride salt. So, this is an example of a reversible reaction.

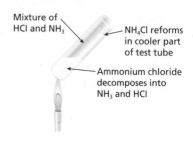

Mixture of HCl and NH_3

NH_4Cl reforms in cooler part of test tube

Ammonium chloride decomposes into NH_3 and HCl

The chemical equations for this reaction are:

$$\text{ammonium chloride} \underset{\text{cool}}{\overset{\text{heat}}{\rightleftharpoons}} \text{ammonia} + \text{hydrogen chloride}$$

$$NH_4Cl(s) \underset{\text{cool}}{\overset{\text{heat}}{\rightleftharpoons}} NH_3(g) + HCl(g)$$

Equilibrium

If reversible reactions happened in a closed system, like a conical flask with a bung in, then:
- no substances can enter
- no substances can leave
- energy can move into and out of the system
- the system will reach equilibrium.

Catalysts increase the rate of the forward and backward reactions by the same amount. This means that catalysts allow reversible reactions in a closed system to get to equilibrium quicker.

At equilibrium:
- the concentration of all the substances is constant
- rate of forward reaction = rate of reverse reaction.

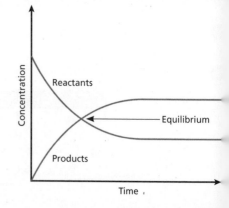

Reversible reactions

Reversible reactions

1 Define 'reversible reaction'.

_____

_____

_____

2 **a)** What is the condition which causes thermal decomposition of ammonium chloride?

_____

b) What is the condition which causes the neutralisation reaction between ammonia and hydrogen chloride gases?

_____

c) The thermal decomposition of ammonium chloride requires 176 kJ/mol. How much energy is released in the neutralisation reaction between ammonia and hydrogen chloride gas?

_____

3 Draw lines to join the boxes.

| The forward reaction of a reversible reaction is exothermic. | | The backward reaction is exothermic. |

| The forward reaction of a reversible reaction is endothermic. | | The backward reaction is endothermic. |

4 What do you observe when hydrated copper(II) sulfate is heated?

_____

Equilibrium

5 **a)** What is needed for a chemical reaction to be at equilibrium?

_____

b) How do the rates of the forward and reverse reactions compare when a chemical reaction is at equilibrium?

_____

c) How do the concentrations of the substances compare when a chemical reaction is at equilibrium?

_____

6 What effect does a catalyst have on a reversible reaction in a closed system?

_____

_____

The effect of changing conditions on reversible reactions

Le Chatelier's Principle

Conditions for an **equilibrium** reaction can be changed and this causes a change in the relative amounts of the substances. The effect of changing the conditions can be predicted using **Le Chatelier's Principle**. It states that for a **reversible reaction** at equilibrium, the system will oppose any change to the conditions.

> Catalysts have no effect on the position of equilibrium.

Changing conditions

Concentration	Temperature	Pressure
If the concentration of one substance is increased, the reaction will move the position of equilibrium to ensure that the concentration returns back to the original level.	If the temperature is changed, the equilibrium system will favour the reaction to return the temperature back to the original level.	Changes in **pressure** only affect an equilibrium where the substances are in the gas phase.

If you increase the concentration of a **reactant**, the system will oppose the change by increasing the rate of the forward reaction. This lowers the concentration of the reactants and increases the concentration of the **products**.

If temperature is increased:
- the system favours the **endothermic** reaction and increases the rate of this reaction
- the relative amount of product increases for a reaction that is endothermic in the forward direction
- the relative amount of product decreases for a reaction that is **exothermic** in the forward direction.

- If pressure increases, the system will increase the rate of reaction for the reaction that has the least amount of gas.

Increase in pressure

Equilibrium shifts to right
More B is converted to A

Increase in reactant

Equilibrium shifts to right
More A is converted to B

If temperature is decreased:
- the system favours the exothermic reaction and increases the rate of this reaction
- the relative amount of product decreases for a reaction that is endothermic in the forward direction
- the relative amount of product increases for a reaction that is exothermic in the forward direction.

- If pressure decreases, the system will increase the rate of reaction which has the most amount of gas.

Decrease in pressure

Equilibrium shifts to left
More B is made

If you increase the concentration of a **product**, the system will oppose the change by increasing the rate of the backward reaction. This lowers the concentration of the products and increases the concentration of the **reactants**.

> Changing pressure on an equilibrium system with no substances in the gas phase will have no effect on the rate of reaction or position of equilibrium.

Increase in product

Equilibrium shifts to left
B is converted to A

The effect of changing conditions on reversible reactions

Le Chatelier's Principle

1 What can Le Chatelier's Principle be used for?

....

....

....

2 What does Le Chatelier's Principle state?

....

....

Changing conditions

3 Complete the sentences by filling in the gaps.

The amount of product .... when more products are added to a

reversible reaction at equilibrium. The amount of product .... when

more reactants are added to a reversible reaction at equilibrium.

The amount of a product .... if the temperature is increased for a

reversible reaction at equilibrium, whose forward reaction is exothermic. The amount of a product

.... if the temperature is increased for a reversible reaction at

equilibrium, whose forward reaction is endothermic.

4 a) What is the effect on the position of equilibrium if the temperature is increased for a reversible reaction at equilibrium?

....

b) What is the effect on the position of equilibrium if the temperature is decreased for a reversible reaction at equilibrium?

....

5 a) What is the effect on the position of equilibrium if the pressure is increased for a gas phase reversible reaction at equilibrium?

....

b) What is the effect on the position of equilibrium if the pressure is decreased for a gas phase reversible reaction at equilibrium?

....

c) What is the effect on the position of equilibrium if the pressure is increased for a reversible reaction at equilibrium with no substances that are gases?

....

Fractional distillation

Crude oil is found in rocks as the remains of an ancient **biomass** consisting mainly of plankton that was buried in mud. It is a:
- **mixture** of **hydrocarbons**
- **fossil fuel**
- **finite** resource.

Fractional distillation is used to separate crude oil into useful fractions. Each fraction is a mixture of **hydrocarbons** with similar **boiling points**.

- It can be done in a lab – a thermometer reads the temperature of the boiling point of the fraction that is being collected. When the temperature shoots up, the collecting flask is changed and the next fraction is collected. This is repeated until all the fractions have been collected.
- In industry, the crude oil is pumped into a furnace and evaporates. The vapour enters the **fractionating column** and the different fractions condense at different heights.

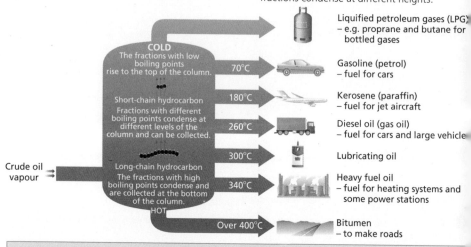

COLD
The fractions with low boiling points rise to the top of the column.

Short-chain hydrocarbon
Fractions with different boiling points condense at different levels of the column and can be collected.

Long-chain hydrocarbon
The fractions with high boiling points condense and are collected at the bottom of the column.
HOT

Crude oil vapour

70°C — Liquified petroleum gases (LPG) – e.g. proprane and butane for bottled gases

70°C — Gasoline (petrol) – fuel for cars

180°C — Kerosene (paraffin) – fuel for jet aircraft

260°C — Diesel oil (gas oil) – fuel for cars and large vehicle

300°C — Lubricating oil

340°C — Heavy fuel oil – fuel for heating systems and some power stations

Over 400°C — Bitumen – to make roads

> The fractions of crude oil are processed by the petrochemical industry to make fuels and materials like solvents, lubricants, polymers and detergents.

Hydrocarbons

Hydrocarbons contain only hydrogen and carbon atoms. The bigger the hydrocarbon:
- the higher its boiling point
- the more **viscous** (thicker) it is
- the less easily **flammable** it is.

Hydrocarbons are often used as fuels and are combusted when used. When **complete combustion** happens, the atoms are fully oxidised and only water and carbon dioxide are produced.

The general equation for this reaction is:

hydrocarbon + oxygen → carbon dioxide + water

Most hydrocarbons in crude oil are **alkanes** with the **general formula** C_nH_{2n+2}.

Alkanes are:
- a **homologous series** (family of chemicals)
- **saturated** (contain only single covalent bonds)
- hydrocarbons.

$H-\overset{\displaystyle H}{\underset{\displaystyle H}{C}}-H$	$H-\overset{\displaystyle H}{\underset{\displaystyle H}{C}}-\overset{\displaystyle H}{\underset{\displaystyle H}{C}}-H$	$H-\overset{H}{\underset{H}{C}}-\overset{H}{\underset{H}{C}}-\overset{H}{\underset{H}{C}}-H$	$H-\overset{H}{\underset{H}{C}}-\overset{H}{\underset{H}{C}}-\overset{H}{\underset{H}{C}}-\overset{H}{\underset{H}{C}}-H$
The simplest alkane, **methane**, CH_4, is made up of 4 hydrogen atoms and 1 carbon atom.	**Ethane, C_2H_6** A molecule made up of 2 carbon atoms and 6 hydrogen atoms.	**Propane, C_3H_8** A molecule made up of 3 carbon atoms and 8 hydrogen atoms.	**Butane, C_4H_{10}** A molecule made up of 4 carbon atoms and 10 hydrogen atoms.

7 Crude oil

Fractional distillation

1 a) What is crude oil made from?

....

b) Where is crude oil found?

....

2 Describe a fraction of crude oil.

....

....

3 Name the **two** physical changes that happen to crude oil to separate it using fractional distillation.

....

Hydrocarbons

4 Which elements are found in hydrocarbons? Circle the correct answers.

Hydrogen	**Helium**	**Oxygen**	**Carbon**	**Calcium**

5 How does the size of the molecule affect flammability?

....

....

6 Describe what happens in complete combustion.

....

7 What type of bonds are found in alkanes?

....

8 Tick the correct equation for the complete combustion of a hydrocarbon.

hydrocarbon + carbon dioxide ➔ oxygen + water ☐

hydrocarbon + oxygen ➔ carbon dioxide + water ☐

hydrocarbon + nitrogen ➔ carbon monoxide + water ☐

hydrocarbon + carbon ➔ carbon dioxide + water ☐

7 Cracking

Cracking

Cracking is the breaking down of long-chain hydrocarbons to make smaller, more useful **hydrocarbons**. This is an **endothermic decomposition** reaction.

The products of cracking are a mixture of:
- **alkanes** – useful for fuels
- **alkenes** – useful for the chemical industry to make products like **polymers**.

Cracking can be done on a small scale in the school lab. The long-chain hydrocarbons are heated until they vapourise and are passed over a catalyst which causes the chemical reaction.

There are two main industrial methods used for cracking:
- **Catalytic cracking** – with a catalyst (e.g. zeolite, aluminium oxide and/or silicon dioxide) and temperatures of 550°C. Very useful for making petrol.
- **Steam cracking** – no catalysts, uses steam, high pressure and temperatures of over 800°C. Makes more alkenes, which are used in the petrochemical industry to make polymers.

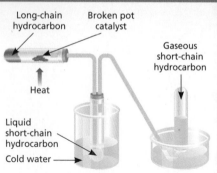

Long-chain hydrocarbon / Broken pot catalyst / Gaseous short-chain hydrocarbon / Heat / Liquid short-chain hydrocarbon / Cold water

long-chain alkane	heat + catalyst →	short-chain alkane + alkene
decane	→	octane + ethene
$C_{10}H_{22}$	→	$C_8H_{18} + C_2H_4$

When crude oil is separated into fractions, there are more long-chain hydrocarbons than are needed. Cracking makes the long-chain hydrocarbons into smaller, more useful hydrocarbons.

Alkenes

Cracking produces a different type of organic molecule with the **general formula** C_nH_{2n}.

Alkenes:
- are a **homologous series**
- are **unsaturated** (contain two fewer hydrogen atoms than the alkane, with the same number of carbon atoms)
- have the **functional group** C=C
- are **hydrocarbons**.

Alkanes and alkenes can be distinguished using a simple laboratory test using bromine water (Br_2(aq)).
- alkanes = no colour change
- alkenes = colour change from orange / brown to colourless.

Modern life depends on many useful materials made by the petrochemical industry, e.g. solvents, lubricants, polymers, detergents. Alkenes are the starting materials for polymers and many of these other substances.

Alkene	Ethene, C_2H_4	Propene, C_3H_6	Butene, C_4H_8	Pentene, C_5H_{10}
Displayed formula	(displayed formula)	(displayed formula)	(displayed formula)	(displayed formula)

Reactions of hydrocarbons

Hydrocarbons contain only carbon and hydrogen atoms. They can be oxidised by:
- complete combustion – excess oxygen and only carbon dioxide and water are made

- incomplete combustion – limited oxygen and a mixture of carbon (soot), carbon dioxide and carbon monoxide are made.

Cracking

1 Define the term 'cracking'.

_____

_____

2 What is made during cracking?

_____

3 What happens to the long-chain hydrocarbon when it is cracked in the lab?

_____

_____

_____

4 **a)** What are the conditions for catalytic cracking?

_____

b) What is catalytic cracking mainly used to make?

_____

Alkenes

5 **a)** Why are alkenes classified as hydrocarbons?

_____

b) Why are alkenes described as unsaturated?

_____

_____

_____

c) What is the functional group in alkenes?

_____

6 Which chemical can be used to distinguish between alkanes and alkenes?

_____

7 What is the molecular formula of propene?

_____

Reactions of hydrocarbons

8 **a)** What are the products of complete combustion of a hydrocarbon?

_____

b) What are the products of incomplete combustion of a hydrocarbon?

_____

⑧ Pure substances

Everyday pure substances and chemically pure substances

In everyday language, substances like milk and water can be described as **pure**. In this context we mean that milk is in its natural state with nothing added, and water is in its natural state with nothing added (**unadulterated**).

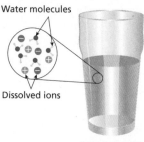

Water molecules

Dissolved ions

Everyday water

In chemistry, milk is not a pure substance as it is a mixture of substances. Water is not a pure substance either, as it is a mixture of substances. In chemistry there is a specific meaning for a pure substance:
- a single **element** or **compound**
- not mixed with any other substance.

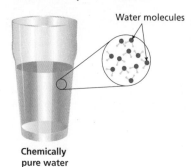

Water molecules

Chemically pure water

Pure substances have specific **melting** and **boiling points**, whereas mixtures will change state over a range of temperatures.

Every pure substance has a unique and precise melting and boiling point, which can be used to identify it. For example, a colourless liquid with a melting point of 0°C and a boiling point of 100°C would be pure water; a liquid with a melting point of –114°C and a boiling point of 78°C would be pure ethanol.

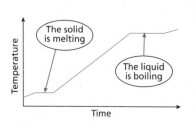

The solid is melting

The liquid is boiling

Temperature

Time

Temperature–time graph for heating a substance

Formulations

A formulation is:
- a **mixture**
- not pure
- designed to be a useful product.

In a formulation, the **components** of the mixture are chosen for a particular purpose. These components are carefully measured and mixed together to make a product with the desired properties.

Use of formulations include:
- fuels
- cleaning agents
- paints
- medicines
- alloys
- foods
- fertilisers.

(8) Pure substances

Everyday pure substances and chemically pure substances

1 What is meant by 'pure' in everyday language?

_____

2 What is meant by 'pure' in terms of chemistry?

_____

3 What particle(s) are found in chemically pure water?

_____

4 Explain why milk is not chemically pure.

_____

_____

5 What substances can be chemically pure?

_____

_____

Formulations

6 **a)** What is a formulation?

_____

b) Why is a formulation not chemically pure?

_____

7 Give **one** use of formulations.

_____

8 Why is a formulation also a type of mixture?

_____

_____

9 How are components of a formulation chosen?

_____

_____

_____

8 Chromatography

Chromatography

Chromatography is a separating technique used to:
- determine if a substance is pure or impure
- separate mixtures
- identify substances.

The process of chromatography is as follows:
- The substance which is being analysed – like an ink or a dye – is put on absorbent paper. This is the **stationary phase** as the paper doesn't move.
- The edge of the paper is placed into a solvent that moves up through the paper. This is the **mobile phase**.
- The different parts of the mixture are attracted to the paper and solvent by different amounts and this causes them to separate.

Paper chromatography

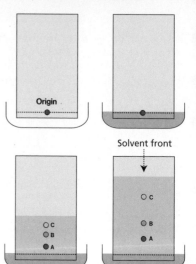

Interpreting a chromatogram

One spot on a **chromatogram** shows that a substance is pure. Multiple dots on a chromatogram show how many components there are in the mixture. The **retention factor**, R_f, value can be used to determine the identity of the substance.

The R_f value is:
- the **ratio** of the distance moved by a substance (centre of spot from origin) to the distance moved by the **solvent**
- different when different solvents (mobile phases) are used
- used to determine an unknown substance.

This chromatogram shows a pure substance as there is only one dot in the chromatogram.

The R_f value can calculated by:

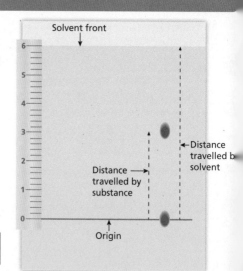

$$R_f \text{ value} = \frac{\text{distance travelled by substance}}{\text{distance travelled by solvent}} = \frac{3}{6} = 0.5$$

Substances can be identified using reference samples on the same chromatogram. The unknown and known substances are used to create the same chromatogram. Any dots that are in line with each other are the same substance.

8 Chromatography

Chromatography

1 The diagram shows the process of chromatography.

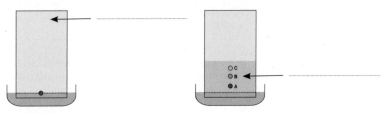

a) Label the mobile phase on the diagram.

b) Label the stationary phase on the diagram.

2 Explain how chromatography works.

Interpreting a chromatogram

3 What is a chromatogram?

4 Complete the sentences by filling in the gaps.

If only one dot appears on the chromatogram, the substance is ....

If more than one dot appears on the chromatogram, the substance is

.... or a ....

5 What is R_f?

6 What is the solvent front?

7 **a)** How do you calculate R_f?

b) What information does the R_f value give you?

Identifying common gases

Producing gas

In a **chemical reaction**, if a gas is produced you may notice:
- **effervescence** (seeing bubbles and hearing fizzing)
- the **mass** decreases as gas is lost to the atmosphere.

Gases can be collected in different ways:

Gases that do not easily dissolve in water, like hydrogen, can be collected by **displacement** of water.	Gases that are denser than air, like carbon dioxide, can be collected by **downward delivery**.	Gases that are less dense than air, like hydrogen, can be collected by **upward delivery**.
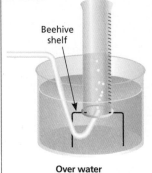 Beehive shelf		
Over water	**Downward delivery (upward displacement of air)**	**Upward delivery (downward displacement of air)**

All gases can be collected and measured using a gas syringe.

Gas tests

If a gas is hydrogen, $H_2(g)$: • Test: use a lighted splint • Result: hear a squeaky pop  Pop	If a gas is carbon dioxide, $CO_2(g)$: • Test: add limewater to the gas and shake • Result: limewater changes from colourless to cloudy CO_2 gas Limewater 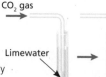
If a gas is oxygen, $O_2(g)$: • Test: use a glowing splint • Result: relights 	If a gas is chlorine, $Cl_2(g)$: • Test: put a piece of damp litmus paper into the gas • Result: turns white (bleaches)

⑧ Identifying common gases

Producing gas

1 What **two** observations are described by the term effervescence?

....

2 Explain why mass might appear to drop when a gas is made in a chemical reaction.

....

....

3 Draw lines to link the method of gas collection with when it would be used.

Method of gas collection	When it would be used
Displacement of water	When the gas is denser than air
Downward delivery	When the gas is less dense than air
Upward delivery	When the gas has low solubility in water

4 **a)** Give an example of a gas that can be collected by displacement of water.

....

b) Give an example of a gas that can be collected by downward delivery.

....

c) Give an example of a gas that can be collected by upward delivery.

....

Gas tests

5 How do you test for hydrogen gas?

....

6 What happens to a glowing splint if oxygen is present?

....

7 Which substance is used to test for carbon dioxide?

....

8 What indicator is used to test for chlorine?

....

9 What happens to indicator paper when it is put into chlorine gas?

....

The development of the Earth's atmosphere

Air and the development of the atmosphere

The envelope of gas around our planet is the **atmosphere**. **Air** is the **mixture** of gases found in the atmosphere and has been the same for about the last 200 million years. Dry air is made of:

- 80% ($\frac{4}{5}$) nitrogen, N_2
- 20% ($\frac{1}{5}$) oxygen, O_2
- trace amounts of carbon dioxide, water vapour and **noble gases**.

Evidence for the development of the atmosphere is limited because of the huge timescale. We are reliant on evidence from rocks, ice cores and space exploration to make hypotheses and plausible theories.

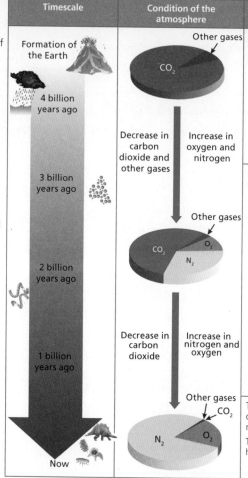

Timescale	Condition of the atmosphere	Key factors and events that shaped the atmosphere
Formation of the Earth 4 billion years ago	Other gases, CO_2	Intense volcanic activity releases: – mainly carbon dioxide (like the atmospheres of Mars and Venus today) – water vapour (which condenses to form the oceans), small proportions of methane and ammonia – nitrogen (which gradually built up in the atmosphere).
3 billion years ago	Decrease in carbon dioxide and other gases / Increase in oxygen and nitrogen	Green plants and algae evolve and: – carbon dioxide is reduced as plants take it in and give out oxygen – microorganisms that can't tolerate oxygen are killed off – carbon from carbon dioxide becomes locked up in sedimentary rocks formed from the shells and skeletons of marine organisms – other gases react with oxygen to release nitrogen – nitrogen is also produced by bacteria removing nitrates from decaying plant material.
2 billion years ago	Other gases, CO_2, O_2, N_2	
1 billion years ago	Decrease in carbon dioxide / Increase in nitrogen and oxygen	There is now about 20% oxygen and about 80% nitrogen in the atmosphere. The amount of carbon dioxide has decreased significantly.
Now	Other gases, CO_2, N_2, O_2	

Oxygen appeared in the early atmosphere about 2.7 billion years ago when **algae** developed. These organisms, later joined by green plants, used **photosynthesis** to obtain the energy from the sun that they needed for life:

carbon dioxide + water → glucose + oxygen
$$6CO_2 + 6H_2O \rightarrow C_6H_{12}O_6 + 6O_2$$

The oxygen in the atmosphere increased over the next billion years until animal life could evolve.

Levels of carbon dioxide in the early atmosphere reduced because of:

- Photosynthesis
- The formation of **carbonate rocks** like limestone. Carbon dioxide dissolved into ocean water and made **soluble carbonates**. These **precipitate** out to make sedimentary rocks, which store carbon dioxide for a very long time.
- The formation of **fossil fuels**, made from ancient **biomass** locked underground, which chemically changes into **hydrocarbons** and locks away some of the carbon from the atmosphere. The stored carbon is then released when the fossil fuels are **combusted** as they are used.

The development of the Earth's atmosphere

Air and the development of the atmosphere

1 What is the atmosphere?

_____

_____

2 **a)** Which gas is the main component of dry air?

_____

b) What (approximate) percentage of dry air is oxygen?

_____

c) What (approximate) fraction of dry air is nitrogen?

_____

3 Which planets in our solar system have a similar atmosphere today as early Earth had?

Circle the answers.

Venus	**Mercury**	**Uranus**	**Mars**	**Jupiter**

4 Where did the gases in the early atmosphere come from?

_____

5 Which organism produced the first oxygen in the atmosphere?

_____

6 What is the name of the process that made oxygen in the atmosphere?

_____

7 What type of rock is limestone?

_____

8 Give **three** ways that carbon dioxide levels were reduced from the early atmosphere.

_____

_____

_____

The greenhouse effect and climate change

Greenhouse gases and the greenhouse effect

Gases in the Earth's **atmosphere** absorb reflected **long-wave radiation** from the Earth's surface. They trap heat energy and keep the average world temperature 14–16°C. These **greenhouse gases** include: water vapour, $H_2O(g)$; carbon dioxide, $CO_2(g)$ and methane, $CH_4(g)$.

The **greenhouse effect** is the natural process in the atmosphere that keeps Earth's average temperature high enough to support life.

Since the industrial revolution, where **combustion** of **fossil fuels** was large scale, the proportion of greenhouse gases has been increasing, which has led to **global warming** – a rise in the average temperature of Earth beyond the natural level.

This is caused by:
- carbon dioxide increasing through combustion of fossil fuels

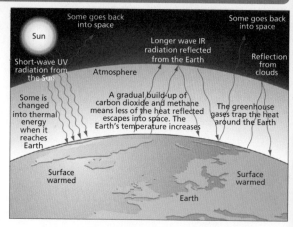

- methane increasing by animal farming, rice paddies, landfills and thawing of permafrost.

Evidence is collected by scientists, which is **peer reviewed** and has been accepted as **accurate**. Scientists have **concluded** that increasing the concentration of greenhouse gases in the atmosphere is causing global warming and leading to **climate change**.

Global climate change and carbon footprint

The effects of climate change include:
- changes in weather patterns – increased flooding, drought and water shortages in summer
- an increase in natural disasters – more powerful and frequent tropical storms, increased number and severity of heatwaves and wildfires
- melting of ice caps – increased sea levels and flooding of low-lying land
- changes to habitats – reduction in **biodiversity**, and industries will need to change (e.g. no snow in ski resorts, different crops needing to be grown by farmers)
- an increase in some diseases – malaria risk in more countries.

Adapting to the new climate may include:
- relocation of people
- flood defence schemes
- developing new technologies including low carbon fuels, more energy-efficient devices and bioengineering crops to withstand extreme weather
- reduction of greenhouse gases by using carbon capture technologies and planting forests

- education of the population
- international treaties.

The total amount of greenhouse gases over the full life cycle of a product, service or event is the **carbon footprint**. This is measured in carbon dioxide equivalent (CO_2e). It can be reduced by:
- extending the lifespan of the product by reusing or recycling
- using renewable energy resources for manufacturing and transportation
- reducing energy used by using local resources and lightweight packaging
- reducing energy loss by insulating or adding lubricants to reduce friction in machines.

It is difficult for people to change their habits but this can be encouraged by laws and financial incentives from government.

> It is difficult to make an accurate model of climate change as simplifications must be used. The media use opinion and speculation as well as the outcome of models, which may lead to bias.

Greenhouse gases and the greenhouse effect

1 Name the **three** main greenhouse gases in our atmosphere.

2 Why is it important that greenhouse gases are in our atmosphere?

3 Explain what humans are doing to the proportion of greenhouse gases in our atmosphere and how they are doing this.

4 What is global warming?

Global climate change and carbon footprint

5 What is climate change?

6 Explain why sea levels might rise.

7 What is a carbon footprint?

8 What is the unit that carbon footprint is measured in?

9 Pollution from combustion

Combustion of fuels

Chemical **fuels** are **combusted** when they are used, and are a major source of atmospheric **pollution**. **Fossil fuels** like coal, oil and natural gas contain **hydrocarbons** (which are the fuel) but also sulfur impurities. When these fuels are combusted they form a **mixture** of **products**:

- water
- carbon dioxide
- carbon monoxide
- carbon (soot)
- sulfur dioxide.

When fuels are combusted in enclosed engines, high pressure and heat can cause nitrogen in the air to **oxidise** to form nitrogen oxides, NOx.

Sulfur dioxide and nitrogen oxides are both **acidic** gases and can cause irritation to the human respiratory system.

Carbon monoxide

Carbon monoxide, CO, is:

- a **toxic** gas
- produced when there is **incomplete combustion** of hydrocarbons
- colourless and odourless.

When carbon monoxide is breathed in, it diffuses into the blood and binds to the haemoglobin in the red blood cells stronger than oxygen. This means that the oxygen-carrying capacity of the blood is reduced. The person will experience light-headedness, headaches and confusion. Carbon monoxide poisoning can lead to death.

Acid rain and global dimming

Rainwater has a **pH** of about 5.5. **Acid rain** is caused when acidic gases dissolve into the rainwater and cause the pH of the rain to be lowered. Acid rain is caused by:

- sulfur dioxide forming sulfuric acid, H_2SO_4
- oxides of nitrogen forming nitric acid, HNO_3.

> NOx are oxides of nitrogen and include nitrogen monoxide, NO, and nitrogen dioxide, NO_2. These are the oxides of nitrogen that are most likely to cause air pollution.

Incomplete combustion of hydrocarbons can cause particles of carbon (soot) to be formed. These are released into the air and cause respiratory problems in humans. In the atmosphere, these particles reflect sunlight and cause **global dimming**.

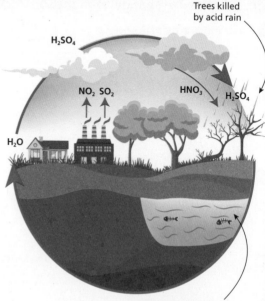

Trees killed by acid rain

H_2SO_4

NO_2 SO_2

HNO_3

H_2SO_4

H_2O

Water becomes acidified, causing fish to die

⑨ Pollution from combustion

Combustion of fuels

1 Which element is an impurity in fuels and can make acidic gases?

....

2 How is NOx made?

....

....

Carbon monoxide

3 **a)** What is the formula of carbon monoxide?

Tick the correct answer.

CO ☐ COx ☐

CO_2 ☐ CO_3 ☐

b) What type of combustion causes carbon monoxide to be made?

....

c) How does carbon monoxide affect the human body?

....

....

Acid rain and global dimming

4 What acid can be formed by sulfur dioxide?

....

5 What acid can be formed by nitrogen oxides?

....

6 Define the term 'acid rain'.

....

....

7 Which atmospheric pollutant causes global dimming?

....

8 Complete the following sentence by filling in the gaps.

Global dimming is caused by .... in the atmosphere, which

.... sunlight.

Resources

Resources are everything in our **environment** that are available to help us satisfy our needs and wants. Humans use the Earth's resources to provide warmth, shelter, food and transport.

Resources are **natural** or **synthetic**.

Natural	Synthetic
Natural resources are used chemically unchanged from the Earth to support life and meet people's needs. They include substances like fossil fuels and conditions like sunlight. **Agriculture** works with natural resources to provide food, timber, clothing and fuels.	**Synthetic materials** are made by chemically changing a natural resource to make a new material. For example, rubber is a natural resource that can be vulcanised by reacting it with sulfur to make it more durable and hardwearing, for use as car tyres. Vulcanised rubber is a synthetic material.

Coal resources
Forest resources
Animal resources
Wind and solar resources
Mineral resources
Oil resources
Water resources
Soil resources

Renewable and finite resources

Resources can be classified by their availability:
- **Renewable resources** are resources that can be replaced as they are being used.
- **Finite resources** are being used up faster than the Earth can replace them.

Finite resources can be found in the Earth, oceans and atmosphere. They are processed to provide energy and materials.

Renewable resources

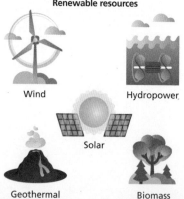

Wind

Hydropower

Solar

Geothermal

Biomass

Finite resources

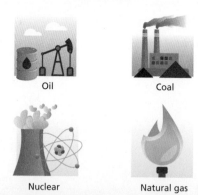

Oil

Coal

Nuclear

Natural gas

The role of chemistry

Chemistry can improve agricultural and industrial processes to create new products and improve **sustainability**. **Sustainable development** ensures that the needs of the people are met today, whilst ensuring that there are enough resources for future generations too.

10 Earth's resources

Resources

1 What do we use resources for?

2 **a)** What is a natural resource?

b) What is a synthetic resource?

3 **a)** Which classification – natural or synthetic – is metal ore?

b) Which classification – natural or synthetic – is a pure metal?

Renewable and finite resources

4 What is a renewable resource?

5 What is a finite resource?

6 Put a tick in the correct column of the table to say whether each resource is renewable or finite.

Resource	Renewable resource	Finite resource
Biomass		
Nuclear		
Coal		
Geothermal		
Wind		
Oil		

The role of chemistry

7 What is sustainable development?

10 Potable water

Essential water

Water, H_2O, is a **natural resource** which is needed for all life. **Potable water** is water that is safe to drink and has sufficiently low levels of:

- **dissolved salts**
- **microbes**.

There are lots of different ways to find or make potable water but it depends on availability of water supplies and the local conditions. Fresh water is used in the United Kingdom to make potable water. **Fresh water** is rain water so it has low levels of dissolved substances and can collect in the ground and in lakes and rivers.

To make potable water in the UK:

- a suitable source of fresh water is chosen, e.g. ground water, rainfall, lakes, rivers, reservoirs, springs
- fresh water is passed through filter beds to remove **insoluble** solids
- the water is **sterilised** using chlorine (Cl_2), ozone (O_3) or UV light to destroy pathogens and prevent disease.

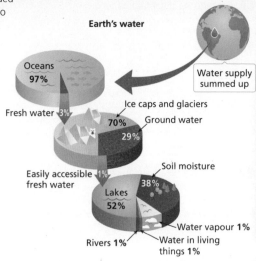

Earth's water

Water supply summed up

Oceans **97%**

Fresh water **3%**

Ice caps and glaciers **70%**

Ground water **29%**

Easily accessible fresh water **1%**

Soil moisture **38%**

Lakes **52%**

Rivers **1%**

Water vapour **1%**

Water in living things **1%**

> Pure water contains only water molecules. Pure water is potable but not all potable water is pure water. For example, tap water in the UK has dissolved substances in it but it is still safe to drink.

Desalination

When fresh water supplies are limited, **desalination** of salty water can be used to make potable water. This can be done by **distillation** or **reverse osmosis**.

Distillation	Reverse osmosis
The salty water is boiled and the water vapour condenses into pure water.	The salty water is put under high pressure and passed through a **membrane** which allows the water molecules through, but very few dissolved ions.
This process is expensive as it requires a lot of energy.	This method makes a large amount of waste water, requires expensive membranes and uses a lot of energy.

Thermometer

Cold water in

Condenser

Distillation adapter

Cork

Distilling flask

Water out

Bunsen burner

Receiving flask

Distillate

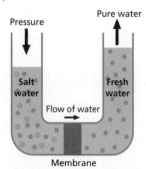

Pressure

Pure water

Salt water

Fresh water

Flow of water

Membrane

10 Potable water

Essential water

1 Describe the difference between pure water and potable water.

2 What is fresh water and where can it be found?

3 What are the **two** main processes in the UK used to make potable water from fresh water?

4 How are insoluble solids removed from fresh water to make potable water in the UK?

5 Suggest an alternative to adding chlorine to UK potable water.

Desalination

6 When is desalination used to make potable water?

7 Why is desalination an expensive way to make potable water?

8 What **two** processes can be used to desalinate water to make potable water?

Sources of waste water

Urban lifestyles, as well as **industrial processes**, produce a large volume of **waste water**. Waste water is water that has been used in the home, business or as part of an industrial or agricultural process. Waste water must be treated before being released into the **environment** in order to prevent **pollution**.

The **urban water cycle** shows how the water resources that people and industry depend on are used.

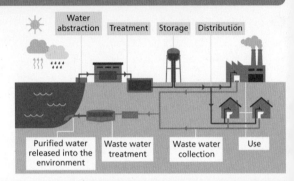

Water abstraction | Treatment | Storage | Distribution

Purified water released into the environment | Waste water treatment | Waste water collection | Use

Treatment of waste water

Different sources of waste water need different **treatments**:

- **Sewage** (water that goes into sewers) and agricultural waste water require removal of **organic matter** and harmful **microbes**.
- Industrial waste water must have harmful chemicals removed and may also require the removal of **organic matter**.

Sewage water is used water from our homes that has been used in baths, sinks, washing machines, dishwashers and toilets. Sewage water in the UK is

treated before the water is returned to rivers or the sea. The treatment includes:

- **Screening** and **grit** removal takes out large insoluble particles.
- **Sedimentation** lets the sewage **sludge** (solid) settle to the bottom and **effluent** (liquid) collect at the top.
- **Bacteria** are used to **anaerobically** digest sewage sludge.
- **Aerobic** bacteria are used to complete a biological treatment of effluent.

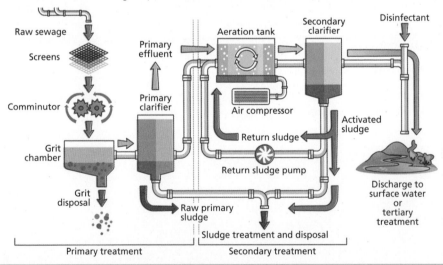

Raw sewage

Screens

Comminutor

Grit chamber

Grit disposal

Primary effluent

Primary clarifier

Raw primary sludge

Aeration tank

Air compressor

Return sludge

Return sludge pump

Secondary clarifier

Disinfectant

Activated sludge

Discharge to surface water or tertiary treatment

Sludge treatment and disposal

Primary treatment | Secondary treatment

Waste water is difficult to make into potable water because several processes are needed as it contains different types of solid waste and pollutants, as well as harmful microbes.

10 Waste water

Sources of waste water

1 Why must waste water be treated?

Treatment of waste water

2 What needs to be removed from sewage waste water?

3 What needs to be removed from waste water that comes from industry?

4 What needs to be removed from agricultural waste water?

5 What does screening and grit removal remove from sewage?

6 **a)** What is effluent?

b) What is sludge?

c) How are effluent and sludge produced from sewage?

7 Why does waste water need to go through the sedimentation process?

8 What digests sewage sludge?

9 What is used to complete the biological treatment of effluent?

10 Alternative methods to extract metals

Ores

Metal **ores** are a **finite natural resource**. Traditional methods of extracting the metals involve:
- finding the metal ore in the Earth's surface
- mining the ore by digging, moving and disposing of large amounts of rock
- processing the ore and reduction of the metal compounds to extract the metals.

Copper (Cu) is an important metal used for electrical wires, water pipes and cooking pots. Copper ores are becoming more difficult to find and are lower-grade (have a lower percentage of copper in them), which means new ways of extracting copper from low-grade ores are needed.

> Unreactive metals like gold are found as uncombined metals in nature. But most metals are found in metal compounds called minerals. When the percentage of metal in the mineral is enough to make it economical to extract the metal, we describe the mineral as an ore.

Metal solutions

Some plants can absorb metal ions as they take in water from their roots. This process can be used in **phytomining** to extract metals from low-grade ores:
- Plants absorb metal ions, e.g copper, Cu^{2+}.
- Plants are harvested.
- Plants are combusted.
- Remaining ash is reacted with acid to make a leachate (solution of metal compounds), e.g. react with sulfuric acid, H_2SO_4, to make copper sulfate, $CuSO_4$.

Plants absorbing metal ions

Soil containing low percentage of copper ore → Plants are burnt in air → Ash containing high percentage of copper compound → Copper metal

Some **bacteria** can absorb metal ions. This process can be used in **bioleaching** to make a **leachate** solution that contains metal compounds.

> Both phytomining and bioleaching are time-consuming, but can be used to conserve metal ores as well as clean up contaminated soils.

Extracting the metal

The metal ions in the leachate solution must be **reduced** to make metal atoms. This can be achieved by **displacement reactions** or **electrolysis**.

Displacement reactions are when a more reactive metal takes the place of the less reactive metal in a compound. The copper containing leachate from phytomining and bioleaching can be reacted with scrap iron to displace copper from its compound:
- Word equation:
  copper sulfate + iron → copper + iron sulfate
- Balanced symbol equation:
  $CuSO_4 + Fe →$
  $Cu + FeSO_4$
- Ionic equation:
  $Cu^{2+} + 2e^- → Cu$

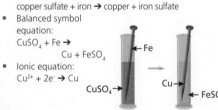

Electrolysis of the leachate can be used to extract the pure metal.

Electrolysis is also used to purify copper made from displacement as pure copper is very useful in electronics as impurities can increase electrical resistance.

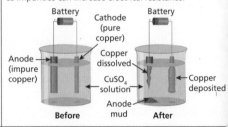

10 Alternative methods to extract metals

Ores

1 What type of natural resource are metal ores?

....

....

2 What are low-grade ores?

....

....

3 Give **one** use of copper metal.

....

Metal solutions

4 **a)** What metal extraction method uses plants?

....

b) What metal extraction method uses bacteria?

....

5 What is a leachate?

....

6 **a)** Give an advantage of using new ways of extracting metals from low-grade ores.

....

....

b) Give a disadvantage of using new ways of extracting metals from low-grade ores.

....

....

Extracting the metal

7 What is a metal displacement reaction?

....

....

8 **a)** What happens to the metal ions in the leachate when they undergo a displacement reaction?

....

b) What happens to the metal ions in the leachate when they undergo an electolysis reaction?

....

(10) Life cycle assessment

Life cycle assessment

The environmental impact of products and services can be assessed using a process called a **life cycle assessment** (LCA).

The LCA considers the environmental impact, including the transport and distribution, of:
- extracting and processing raw materials
- manufacturing and packaging of a product
- use and operation of a product during its lifetime
- disposal at the end of its useful life.

Scientists like to be **objective**, which means that facts are represented without bias because of personal feelings or opinions. The LCA is not purely objective because allocating numerical values to pollutant effects is less straightforward and more **subjective** as they require value **judgements**.

LCAs of different shopping bags can be used to justify which type of bag is best in terms of environmental impact.

Researchers can then use data and models to answer the investigation points, considering each stage of the LCA in turn:
- Stage 1 raw materials – Paper is made from wood, which is sustainable and renewable. Plastic is made from crude oil, which is finite.
- Stage 2 manufacturing – Most energy is needed to make plastic from crude oil. Paper made from trees requires more energy than making recycled paper.
- Stage 3 use – Plastic bags are lighter and require less energy to transport compared to paper. Plastic bags are durable and can be re-used many times; paper bags are often single use.
- Stage 4 disposal – Plastic bags can always be recycled, but if they are put in landfill, they will not biodegrade. Some paper bags can be recycled up to 7 times and will biodegrade in landfill.

Researchers then must arrive at a **judgement** using the LCA. The data suggest that plastic bags have lower impact on the environment than paper bags as long as they are used many times and then recycled. The key is long life of the product and responsible disposal.

Reduce, re-use and recycle

For more **sustainable** living we should:
- **Reduce** the amount of materials we use, e.g. use thinner aluminium foil.
- **Re-use** products, e.g. a glass bottle can be cleaned and sterilised then re-used, or metals can be recycled by melting and recasting or reforming into different products.
- **Recycle** materials, e.g. glass is crushed and melted; metals are melted and recast or reformed.

- Sometimes recycled materials can be added wit new materials, e.g. scrap steel can be added to iron from the blast furnace, reducing the amount of new iron ore needed to make iron.

> Marketing and industry can summarise LCAs to ensure they produce a favourable, pre-determined conclusion. This can lead to supported claims for advertising purposes, which are not objective or comprehensive.

Reduce

Re-use

Recycle

10 Life cycle assessment

Life cycle assessment

1 What does an LCA measure?

2 a) What parts of the LCA are subjective?

b) What parts of the LCA are objective?

c) Why are some parts of the LCA objective?

3 What can happen to a paper bag at the end of its useful life?

4 What is a judgement?

Reduce, re-use and recycle

5 Describe what must be done to a glass bottle before it can be re-used.

6 How is metal recycled?

7 How can glass be recycled?

1 Energy stores and systems

Energy stores

Energy is never created, or lost, it is just **transferred** between energy stores. This is known as the **law of conservation of energy**.

Suppose a change takes place that causes energy to be transferred from one store to another: if the first energy store decreases by 100 J, the second energy store increases by 100 J.

The five main energy stores are:

Kinetic energy store	Thermal energy store	Chemical energy store
If the cyclist pedals faster, his store of kinetic energy increases.	The poker's store of thermal energy increases when it is put into the fire.	Chemical energy is transferred if the battery is connected into a circuit or if the coal is ignited.

Gravitational potential energy store	Elastic potential energy store	
As the skydiver falls, her store of gravitational potential energy decreases.	When the archer fires the arrow, the bow's store of elastic potential energy decreases.	Energy can never be created or destroyed, but it can be transferred between different stores.

Energy transfers

A system is an object or group of objects; when a system changes, there are changes in the way energy is stored. Energy can be transferred between the different energy stores by a force, by heating, by an electric current or by a wave.

Whenever you use a domestic appliance that connects to the mains electricity supply, there is an energy transfer by an electric current.	Whenever a force is applied to an object, mechanical work is done and there is an energy transfer by the force.	If an object emits light or sound, there is an energy transfer by radiation. For example, when a lamp is switched on, there is an energy transfer along the electrical pathway transferring energy into the thermal store and into the radiation pathway.

The sequence of energy transfers when a portable gas stove heats water in an aluminium can is shown. The chemical reaction involving combustion of propane and oxygen results in a transfer of energy.

Chemical energy stored in propane and oxygen → Combustion of propane and oxygen transfers chemical energy to thermal energy → Thermal energy transferred by a temperature difference →

→ Thermal energy store of aluminium can increases

→ Thermal energy store of water increases

→ Thermal energy store of surroundings increases

Energy can be transferred between stores by a force, by heating, by an electric current or by a wave.

1 Energy stores and systems

Energy stores

1 State the law of conservation of energy.

....

....

2 The stem of a pear breaks. The force of gravity acting on the pear causes it to fall from the tree, resulting in a transfer of energy.

a) Name the energy store that is decreasing.

....

b) Name the energy store that is increasing as the pear falls.

....

3 A cyclist applies the brakes on seeing a hazard in the road ahead. The force of friction brings the bike to a stop and raises the temperature of the brakes.

a) Name the energy store that has decreased.

....

b) Name the energy store that has increased.

....

Energy transfers

4 The incomplete flow chart represents energy changes when a lamp is switched on.
Fill in the missing words.

| The lamp battery is a store of .... energy. | → | The lamp is switched on to complete the .... | | Energy is transferred to the lamp bulb by the | → | The thermal energy store of the surroundings increases. |

....

....

5 Describe the energy transfers when a ball is thrown and hits a wall.

....

....

....

....

....

....

1 Changes in energy (1)

Kinetic energy

When an object is in motion, it has energy stored within the **kinetic** store. However, it is not always true to say that the faster an object moves, the greater the energy in the kinetic store. This is because kinetic energy depends on two factors:
- the mass of the object
- the speed at which the object is moving.

The lorry shown below has a mass of 25 000 kg.

The car shown below has a mass of 1200 kg.

Even though the lorry is likely to be moving at a slower speed than the car, it might have more kinetic energy due to its greater mass.

> The energy in the kinetic store of a moving object depends on its mass and its speed.

Calculating kinetic energy

You can calculate the energy in the kinetic store of an object using the following equation:

$$E_k = \frac{1}{2}mv^2$$

where

E_k is the kinetic energy, in joules (J)

m is the mass, in kilograms (kg)

v is the speed, in metres per second (m/s)

Sometimes the mass or the speed of the object will not be known, so you will need to rearrange the equation $E_k = \frac{1}{2}mv^2$ to make the unknown quantity the subject. Remember that when rearranging equations, whatever is done to one side must be done to the other.

A lorry with a mass of 25 000 kg is travelling at 15 m/s. Calculate the energy in the kinetic store of the lorry.

Begin by listing the information known about the lorry.

Mass = 25 000 kg

Speed = 15 m/s

Substitute these values into the equation for kinetic energy.

$E_k = \frac{1}{2}mv^2$

$E_k = \frac{1}{2} \times 25\,000 \times 15^2$

$E_k = 2\,812\,500\,J$

If the mass is unknown		If the speed is unknown	
$E_k = \frac{1}{2}mv^2$		$E_k = \frac{1}{2}mv^2$	
$2E_k = mv^2$	Multiply both sides of the equation by 2	$2E_k = mv^2$	Multiply both sides of the equation by 2
$\frac{2E_k}{v^2} = m$	Divide both sides of the equation by v^2	$\frac{2E_k}{m} = v^2$	Divide both sides of the equation by m
		$\frac{\sqrt{2E_k}}{m} = v$	Take the square root of both sides of the equation

> The energy in the kinetic store of a moving object can be calculated by $E_k = \frac{1}{2}mv^2$

① Changes in energy (1)

Kinetic energy

1 State the **two** factors that determine the energy within the kinetic store of a moving object.

....

....

2 A bus and a car are both moving at a speed of 20 m/s. The bus has a mass of 15 000 kg and the car has a mass of 1100 kg.

Without calculation, state and explain which object will have the greater kinetic energy.

....

....

....

....

....

Calculating kinetic energy

3 A cyclist and their bike have a combined mass of 100 kg and are travelling at a speed of 7 m/s.

Calculate the kinetic energy of the cyclist and bike. Remember to state the unit.

....

4 A train has a mass of 30 000 kg and is travelling at a speed of 56 m/s.

Calculate the kinetic energy of the train in kilojoules.

.... kJ

5 A rugby player runs at a speed of 4 m/s and has 640 J of energy in the kinetic store.

Calculate the mass of the rugby player in kilograms.

.... kg

6 A Boeing 747 aeroplane has a mass of 184 000 kg and has 5 750 000 000 J of energy in its kinetic store.

Calculate the speed of the aeroplane in metres per second.

.... m/s

(1) Changes in energy (2)

Elastic potential energy

When a spring is stretched, it has energy stored within the **elastic potential energy** store. The amount of energy stored depends on two factors:
- how stiff the spring is
- how far the spring extends from its original length – the **extension**.

The stiffness of a spring is given a value, known as the **spring constant**, k. It is defined as the force required to stretch the spring by 1 m.

The elastic potential energy can be calculated using the following equation:

> Elastic potential energy = $\frac{1}{2}$ × spring constant × extension²
>
> $$E_e = \frac{1}{2}ke^2$$
>
> where E_e is the elastic potential energy, in joules (J)
>
> k is the spring constant, in newtons per metre (N/m)
>
> e is the extension, in metres (m)

The spring has a spring constant of 150 N/m and extends by 2 cm.
Calculate the elastic potential energy stored in the spring.

$E_e = \frac{1}{2}ke^2$

$E_e = \frac{1}{2} \times 150 \times 0.02^2$ Remember to convert the extension of the spring into metres.

$E_e = 0.03\,J$

Elastic potential energy depends on the spring constant and the extension of the spring. It can be calculated using the equation $E_e = \frac{1}{2}ke^2$

Gravitational potential energy

When an object is raised above the Earth's surface, it gains **gravitational potential energy**, E_p. The amount of gravitational potential energy gained by an object depends on three factors:
- the **mass** of the object
- the **gravitational field strength**
- the **height** the object is raised to.

The gravitational field strength is the force of gravity per unit mass, measured in newtons per kilogram, N/kg.

The **gravitational potential energy** can be calculated using the following equation:

> Gravitational potential energy = mass × gravitational field strength × height
>
> $$E_p = mgh$$

The summit of Mount Snowdon in Wales is 1085 m above sea level.

Calculate the gravitational potential energy gained by a mountaineer with a mass of 70 kg when they reach the summit of Mount Snowdon. Take the gravitational field strength, g, to be 10 N/kg.

$E_p = mgh$

$E_p = 70 \times 10 \times 1085$

$E_p = 759\ 500\ J = 759.5\ kJ$

Gravitational potential energy depends on the mass, the gravitational field strength and the height. It can be calculated using the equation $E_p = mgh$

① Changes in energy (2)

Elastic potential energy

1 State the **two** factors that elastic potential energy depends on.

_____

_____

2 A spring has a spring constant of 300 N/m. When a force is applied, it extends by 5 cm.

Calculate the elastic potential energy stored in the spring.

_____ J

3 A spring has a spring constant of 500 N/m. When a force is applied, it extends from 5 cm to 25 cm.

Calculate the elastic potential energy stored in the spring.

_____ J

4 A stretched spring stores 20 J of elastic potential energy when it extends by 40 cm.

Calculate the spring constant, k, of the spring in N/m.

_____ N/m

Gravitational potential energy

5 State the **three** factors that gravitational potential energy depends on.

_____

_____

_____

6 A book of mass 0.75 kg is placed on a table that is 1.2 m high.

Calculate the gravitational potential energy gained by the book. Take g to be 10 N/kg.

_____ J

7 A golf ball has a mass of 45 g.

Calculate the gravitational potential energy gained by the golf ball when it is hit 20 m into the air. Take g to be 10 N/kg.

_____ J

8 A mountaineer of mass 75 kg gains 637.5 kJ of gravitational potential energy when they climb a mountain.

Calculate the height of the mountain that they climb. Take g to be 10 N/kg.

_____ m

(1) Energy changes in systems

Specific heat capacity

Have you ever walked on a beach and felt the heat of the sand on your feet? What do you feel as your feet touch the water?

The water feels much cooler than the sand, despite them both having been under the same Sun. This is because sand and water have very different specific heat capacities.

The **specific heat capacity** is a measure of how much energy it takes to raise the temperature of 1 kg of a particular material by 1°C.

> The specific heat capacity is the energy required to raise the temperature of 1 kg of a substance by 1°C.

Calculating changes in energy

The amount of energy required to heat up a substance depends on three factors:
* the mass of the substance being heated
* the specific heat capacity of the substance being heated
* the change in temperature.

You can use the following equation to calculate the amount of energy required to heat up a substance:

$$\Delta E = mc\Delta\theta$$

where

ΔE is the change in thermal energy, in joules (J)

m is the mass, in kilograms (kg)

c is the specific heat capacity, in joules per kilogram per degree Celsius (J/kg°C)

$\Delta\theta$ is the change in temperature, in degrees Celsius (°C)

An aluminium block has a mass of 1.5 kg.

Calculate the energy required to raise the temperature of this block from 15°C to 35°C. The specific heat capacity of aluminium is 900 J/kg°C.

Begin by listing the information known about the block.

Mass = 1.5 kg
Specific heat capacity = 900 J/kg°C
Change in temperature = (35 − 15) = 20°C

Substitute these values into the equation for change in energy.

$\Delta E = mc\Delta\theta$
$\Delta E = 1.5 \times 900 \times 20$
$\Delta E = 27\,000$ J

Sometimes the mass, the change in temperature or the specific heat capacity of the object will not be known, so you will need to rearrange the equation $\Delta E = mc\Delta\theta$ to make the unknown quantity the subject. Remember that when rearranging equations, whatever is done to one side must be done to the other.

If the mass is unknown		If the change in temperature is unknown		If the specific heat capacity is unknown	
$\Delta E = mc\Delta\theta$		$\Delta E = mc\Delta\theta$		$\Delta E = mc\Delta\theta$	
$\dfrac{\Delta E}{(c\Delta\theta)} = m$	Divide both sides of the equation by $(c\Delta\theta)$	$\dfrac{\Delta E}{(mc)} = \Delta\theta$	Divide both sides of the equation by (mc)	$\dfrac{\Delta E}{(m\Delta\theta)} = c$	Divide both sides of the equation by $(m\Delta\theta)$

> The energy needed to heat a substance depends on the mass of the substance, its specific heat capacity and the change in temperature. It is calculated using the equation $\Delta E = mc\Delta\theta$

(1) Energy changes in systems

Specific heat capacity

1 What is meant by the term 'specific heat capacity'?

....

....

2 State the **three** factors that affect the amount of energy needed to heat up a substance.

....

....

....

Calculating changes in energy

3 A copper coin has a mass of 3.5 g. The specific heat capacity of copper is 385 J/kg°C.

Calculate the change in energy of the coin when it is heated from 20°C to 50°C.
Give your answer to 3 significant figures and remember to state the unit.

....

4 Calculate the change in energy when 2 kg of water is heated from 18°C to 100°C.
Give your answer in kilojoules. The specific heat capacity of water is 4200 J/kg°C.

.... kJ

5 An aluminium block is heated from 15°C to 50°C. 78 750 J of energy is supplied to the block.
The specific heat capacity of aluminium is 900 J/kg°C.

Calculate the mass of the aluminium block in kilograms.

.... kg

6 A bar of steel of mass 1.5 kg is heated using 28.35 kJ of energy.
The specific heat capacity of steel is 420 J/kg°C.

Calculate the change in temperature of the steel bar.

.... °C

Power

An electrical appliance transfers energy stored in the mains electrical supply to other energy stores. When you switch on an electric kettle, an electric current flows through its heating element, transferring energy from the mains.

The energy transfer increases the thermal energy stores of the water in the kettle, the body of the kettle and its surroundings.

The **power** of an electrical appliance is the energy it transfers in 1 second.

1 joule of energy transferred in 1 second represents a power of 1 watt, written as 1 W.

Household electrical appliances are labelled with their power. If the label on a hairdryer shows that its power is 800 W, this means that it transfers 800 J of energy from the mains supply every second.

> The power of an electrical appliance is the energy it transfers in 1 second. 1 joule of energy transferred in 1 second represents a power of 1 watt (1 W).

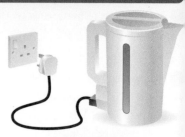

Calculating energy transfer

The energy transferred by a device can be calculated using the following equation:

$$power\ (W) = \frac{energy\ (J)}{time\ (s)}$$

This equation can be rearranged to:
energy (J) = power (W) × time (s)

Work done is equal to energy transferred, so:

$$power\ (W) = \frac{work\ done\ (J)}{time\ (s)}$$

If an 800 W hairdryer is used for 5 minutes, how much energy will it transfer?

First convert 5 minutes into seconds

Time = 5 × 60 = 300 s

Now use the energy equation

Energy transferred = 800 × 300 = 240000 J

Appliances designed to produce thermal energy usually have greater powers. High powers are often given in kilowatts (kW). For example, the power of an iron may be given as 2.4 kW.

> The power can be calculated using the equations:
> $$power\ (W) = \frac{energy\ (J)}{time\ (s)} \quad and$$
> $$power\ (W) = \frac{work\ done\ (J)}{time\ (s)}$$

1 Power and energy transfer

Power

1 Complete the following sentence.

An energy transfer of 1 .... per .... is

equal to a power of 1 ....

2 Two electric motors, A and B, lift the same weight through the same height.

Motor A takes 3 seconds to lift the weight. Motor B takes 10 seconds to lift the same weight through the same height.

Which motor is more powerful? Explain your answer.

....

....

....

Calculating energy transfer

3 An electric immersion heater transfers 10 000 000 J in 20 minutes.

a) Calculate the power of the immersion heater in watts.

.... W

b) Convert your answer in part a) to kilowatts.

.... kW

4 The power rating on an electric kettle is 2500 W.

a) Explain what this means in terms of how quickly it transfers energy.

....

b) The kettle full of water takes 3 minutes to boil.

Calculate how much energy is transferred by the kettle during the 3 minutes it takes to boil the water. Give your answer in kilojoules.

.... kJ

5 The engines in a train have a combined power of 2 MW.

a) Convert the power of the engines into watts. Give your answer in standard form.

.... W

b) Calculate the energy transferred by the train over a 30-minute journey, assuming that the engines run at full power for the duration of the journey. Give your answer in MJ.

.... MJ

(1) Energy transfers in a system

Law of conservation of energy

The **law of conservation of energy** states that energy cannot be created or destroyed. Energy can be transferred between different stores, but the total energy in a system remains constant.

If a model car is pushed and then allowed to come to a stop, there is an energy transfer from the kinetic store of the car to the thermal store of the surroundings as work is done by friction in bringing the car to a stop.

The total energy within this system remains constant. In other words, there is no net change in total energy.

When devices transfer energy, only part of the energy is usefully transferred to where it is wanted and in the form that is wanted. The remaining energy is **transformed** in a non-useful way, mainly as thermal energy. This is known as wasted energy.

For example, a light bulb transforms electrical energy into useful light energy. However, for some light bulbs, most of the energy is wasted as heat energy.

The wasted energy and the useful energy are eventually transferred to their surroundings, which become warmer. It is said that the energy has been **dissipated**.

> Energy cannot be created or destroyed. It can only be transferred from one store to another.

Reducing unwanted energy transfers

It is not possible to completely prevent unwanted energy transfers, but you can reduce them.

Applying oil to a bike chain

Where two surfaces rub together, there is always an energy transfer into the thermal store of the objects. **Lubricating** the surfaces (using oil or graphite) reduces friction and therefore reduces the energy transfer into the thermal store.

Applying oil to a bicycle chain reduces the energy transfer to the thermal store, meaning that more energy is transferred into the kinetic store. This increases the speed for the same energy input.

Cavity wall insulation

Buildings can be **insulated** with **thermal insulation** to reduce unwanted energy transfers. Cavity wall insulation and loft insulation can be used to prevent heat loss through the walls and roofs of houses.

Cavity wall insulation is made of a material containing pockets of trapped air that is injected into the gaps between the interior and exterior walls of houses. This reduces heat loss via conduction and convection.

Loft insulation is made from a material with a low **thermal conductivity**. This means that the material is a poor conductor of heat, therefore insulating the building.

Loft insulation

The rate of cooling of a building depends on both the **thickness** and **thermal conductivity** of the walls. Thick walls, insulated with a material with a low thermal conductivity, have the slowest rate of heat loss.

> Unwanted energy transfers can be reduced, for example through lubrication and the use of thermal insulators. Effective thermal insulators have a low thermal conductivity.

1 Energy transfers in a system

Law of conservation of energy

1 State the law of conservation of energy.

_____

_____

_____

2 When a television is turned on, there are useful and wasted energy transfers.

Identify **one useful** and **one wasted** energy transfer.

_____

_____

_____

Reducing unwanted energy transfers

3 It is important to ensure that the oil in a car engine is always topped up.

Explain why oiling moving parts reduces unwanted energy transfers.

_____

_____

_____

4 State **two** ways in which unwanted energy transfers can be reduced in the home.

_____

_____

5 Thermal conductivity is a measure of the rate at which heat energy passes through a material. The higher the thermal conductivity of a material, the greater the rate of energy transfer through the material.

Use this information to explain why loft insulation is made of a material with low thermal conductivity.

_____

_____

_____

_____

_____

(1) Efficiency

Efficiency and how it is calculated

Devices are designed to waste as little energy as possible. This means that more energy can be usefully transferred.

The **efficiency** of a device is the proportion of energy that is usefully transformed. The greater the proportion of energy that is usefully transformed, the more efficient the device is.

The efficiency of a device can be calculated using the equations below. Note that power is a measure of the rate of energy transfer.

Wasted energy Heat 150 joules/s

Useful energy Light 20 joules/s

Useful energy Sound 30 joules/s

Electrical energy 200 joules/s

If a quarter of the energy supplied to a television is usefully transformed into light and sound, it is only 25% efficient.

$$\text{efficiency} = \frac{\text{useful output energy transfer}}{\text{total input energy transfer}}$$

$$\text{efficiency} = \frac{\text{useful power output}}{\text{total power input}}$$

The total energy supplied to a light bulb is 200 J. The bulb usefully transfers 105 J of energy. Calculate the efficiency of the light bulb.

Total input energy transfer = 200 J
Useful output energy transfer = 105 J

Begin by listing the information known about the energy transfer

Substitute these values into the equation for efficiency

$$\text{efficiency} = \frac{\text{useful output energy transfer}}{\text{total input energy transfer}}$$

$$\text{efficiency} = \frac{105\,\text{J}}{200\,\text{J}}$$

efficiency = 0.525

Notice that efficiency does not have any units. This is because the units of joules (or watts) cancel out on the top and the bottom of the fraction

A hairdryer has a power rating of 2000 W. It usefully transfers energy at a rate of 1400 W. Calculate the percentage efficiency of the hairdryer.

Calculate the efficiency of the hairdryer by $\frac{\text{useful power output}}{\text{total power input}}$ and multiply by 100%

$$\text{percentage efficiency} = \frac{1400}{2000} \times 100\%$$

percentage efficiency = 70%

Rearranging the efficiency equation

If the useful output energy transfer is unknown	If the total input energy transfer is unknown
$\text{efficiency} = \frac{\text{useful output energy transfer}}{\text{total input energy transfer}}$	$\text{efficiency} = \frac{\text{useful output energy transfer}}{\text{total input energy transfer}}$
Multiply both sides of the equation by the total input energy transfer	Multiply both sides of the equation by the total input energy transfer
	efficiency × total input energy transfer = useful output energy transfer
$\text{efficiency} \times \text{total input energy transfer} = \text{useful output energy transfer}$	Divide both sides of the equation by the efficiency
	$\frac{\text{useful output energy transfer}}{\text{efficiency}} = \text{total input energy transfer}$

Increasing the efficiency of a device

Devices can't have an efficiency greater than 1 or 100%. This would mean that more useful energy was transferred by the device than was put into it, breaking the law of conservation of energy.

Energy can be lost due to friction between moving parts, unwanted sound output or by electrical resistance. **Lubricating** moving parts and using thermal **insulation** can reduce the unwanted energy transfers and, therefore, increase the efficiency of a device.

1 Efficiency

Efficiency and how it is calculated

1 What is meant by the 'efficiency' of a device?

2 Explain why the efficiency of a device can never be greater than 1.

3 An electric motor usefully transfers energy at a rate of 6 W. The input power of the motor is 15 W.
Calculate the efficiency of the motor.

4 300 J of energy are supplied to an LED bulb. The bulb usefully transfers 240 J of energy via the radiation pathway.
Calculate the efficiency of the LED bulb.

Rearranging the efficiency equation

5 A television has an efficiency of 0.80 and it usefully transfers 5500 J of energy.
Calculate the total input energy transfer.

.... J

Increasing the efficiency of a device

6 A kettle has an efficiency of 0.75
How could the efficiency of the kettle be increased?

National and global energy resources

Energy resources on Earth

Humans use energy resources for transport, generating electricity and heating.

Some of these resources are **renewable**. This means that they are being, or can be, replenished as they are used. Others are **non-renewable** because they can't be replaced within a lifetime and will eventually run out.

Nuclear fuels such as uranium and plutonium are non-renewable. Nuclear fuel isn't burned like coal, oil or gas to release energy and it isn't classed as a fossil fuel.

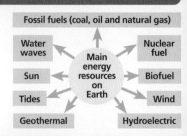

Main energy resources on Earth

- Fossil fuels (coal, oil and natural gas)
- Water waves
- Nuclear fuel
- Sun
- Biofuel
- Tides
- Wind
- Geothermal
- Hydroelectric

Non-renewable energy resources

Source	Advantages	Disadvantages
Coal	• Relatively cheap and easy to obtain. • Coal-fired power stations have a relatively quick start-up time. • There may be over a century's worth of coal left.	• Burning produces CO_2 and SO_2. • Produces more CO_2 per unit of energy than oil or gas does. Removing the SO_2 is costly. • SO_2 causes **acid rain**.
Oil	• Relatively easy to find, though the price is variable. • Oil-fired power stations are flexible in meeting demand. • There is enough left for the short–medium term.	• Burning produces CO_2 and SO_2. • Produces more CO_2 per unit of energy than gas. • Tankers pose the risk of **spillage** and **pollution**.
Gas	• Gas-fired power stations have the quickest start-up time. • There is enough left for the short–medium term. • Doesn't produce SO_2.	• Burning produces CO_2 (although it produces less per unit of energy than coal or oil). • Expensive pipelines and networks are often required to transport it.
Nuclear	• Cost and rate of fuel is relatively low. • Nuclear power stations are flexible in meeting demand. • Doesn't produce CO_2 or SO_2.	• Waste can stay **dangerously radioactive** for thousands of years. • Building and decommissioning is costly. • Longest start-up time.

Renewable energy resources

Many renewable energy sources are 'powered' by the Moon or Sun. The gravitational pull of the Moon creates tides, while the Sun causes evaporation (resulting in rain and flowing water) and convection currents (resulting in winds and waves).

Source	Advantages	Disadvantages
Wind	• No fuel and little maintenance required. • No pollutant gases produced. • Can be built offshore.	• Turbines cause noise and visual pollution. • Not very flexible in meeting demand. • High capital outlay needed to build them.
Tidal and Waves	• No fuel required. • No pollutant gases required. • Barrage water can be released when electricity demand is high.	• They are unsightly, a hazard to shipping and destroy habitats. • Variations of tides and waves affect output. • High capital outlay needed to build them.
Hydro-electric	• Fast start-up time. • No pollutant gases produced. • Water can be pumped back to the reservoir when electricity demand is low.	• Often involves damming upland valleys. • There must be adequate rainfall in the region where the reservoir is. • Very high initial capital outlay needed.
Solar	• Can produce electricity in remote locations. • No pollutant gases produced.	• Dependent on intensity of light. • High cost per unit of electricity produced.

National and global energy resources

Energy resources on Earth

1 State **three** energy resources that are available on Earth.

_____

_____

_____

2 Give **two** uses of energy resources.

_____

_____

3 Name the **three** fossil fuels.

_____

_____

Non-renewable energy resources

4 What is meant by a 'non-renewable' energy resource?

_____

_____

_____

5 Give **one advantage** and **one disadvantage** of using coal as an energy resource.

Advantage: _____

_____

Disadvantage: _____

_____

Renewable energy resources

6 What is meant by a 'renewable' energy resource? Give **one** example of such a resource.

_____

_____

_____

7 Give **one advantage** and **one disadvantage** of using wind as an energy resource.

Advantage: _____

_____

Disadvantage: _____

_____

2 Standard circuit diagram symbols

Electric circuits

An electric circuit is a complete path around which an electric current can flow.

A circuit contains:
- a cell or a battery of cells
- at least one device that can transfer energy, e.g. a lamp
- wires, called connecting leads, which join the cell to the components of the circuit
- a switch (usually included).

> An electric current needs a complete circuit in order to flow. A break in the circuit will stop the current from flowing.

Circuit diagram symbols

When drawing electric circuits, straight lines are used to represent the electrical wires and standard symbols are used instead of drawing each component.

For example:

The circuit below has been set up in a laboratory.

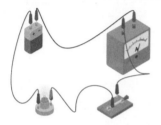

The circuit diagram for this would be:

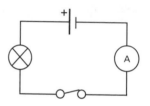

This table shows the symbols that you need to be able to recognise and draw.

Switch (open)	Switch (closed)	Cell	Battery
Thermistor	Diode	Resistor	Variable resistor
Lamp	LED	LDR	
Fuse	Voltmeter	Ammeter	

② Standard circuit diagram symbols

Electric circuits

1 What is meant by an 'electric circuit'?

....

....

2 Write down **four** components of an electric circuit.

....

....

....

....

Circuit diagram symbols

3 Draw the circuit symbols for the following components.

   a) Battery

   b) Variable resistor

   c) Voltmeter

   d) LDR

4 Use standard circuit diagram symbols to draw this circuit.

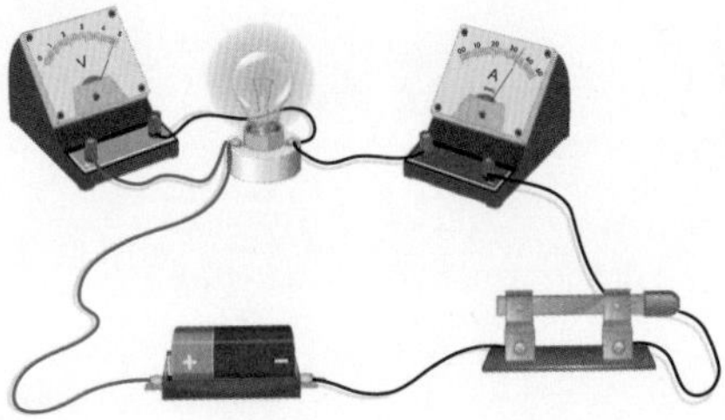

5 Use standard circuit diagram symbols to draw this circuit.

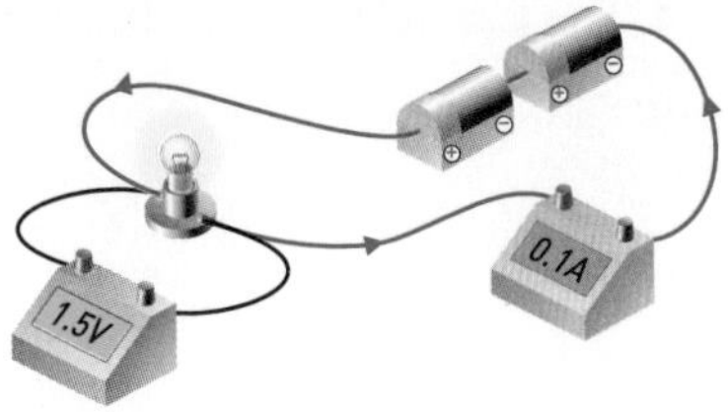

② Electrical charge and current

Electrical charge and current

Electrical current is the flow of electric charge.

The magnitude of the electric current is a measure of the rate of flow of electric charge around the circuit. The greater the current, the faster the flow of electric charge around the circuit.

Calculating electric charge

You can calculate the electric charge and current using the following equation:

$$Q = It$$

where

Q is the electric charge, in coulombs (C)

I is the electric current, in amperes (or amps) (A)

t is the time, in seconds (s)

A current of 10 A flows through a circuit for 5 minutes.

Calculate the electric charge in the circuit.

Begin by listing the information known about the circuit.

Current = 10 A

Time = 5 minutes = 300 seconds

Substitute these values into the equation for electric charge.

$Q = It$

$Q = 10 \times 300$

$Q = 3000\,C$

Rearranging the equation

Sometimes the current or the time will not be known, so you will need to rearrange the equation $Q = It$ to make the unknown quantity the subject. Remember that when rearranging equations, whatever is done to one side must be done to the other.

If the current is unknown		If the time is unknown	
$Q = It$		$Q = It$	
$\frac{Q}{t} = I$	Divide both sides of the equation by time.	$\frac{Q}{I} = t$	Divide both sides of the equation by current.

A circuit transfers 5500 C of electric charge when a current of 5 A flows through it.

Calculate the time the circuit was switched on for.

Give your answer in minutes and seconds.

Begin by listing the information known about the circuit.

Electric charge = 5500 C
Current = 5 A
Resistance = 20 ohms

Substitute these values into the equation for electric current.

$Q = It$

$t = \frac{Q}{I}$

$t = 1100$ seconds = 18 minutes 20 seconds

The electrical charge, current and time are related by the equation $Q = It$

② Electrical charge and current

Electrical charge and current

1 What is meant by 'electric current'?

....

Calculating electric charge

2 In words, write the equation that links current, electric charge and time.

....

3 A current of 3 A flows through a circuit for 10 minutes.

Calculate the electric charge transferred by the circuit in this time.

.... C

4 A circuit is switched on for 8 minutes.
A current of 15 A flows through it.

How much electric charge is transferred by the circuit?

.... C

Rearranging the equation

5 A circuit with a current of 1.5 A transfers 2000 C of electric charge.

How long is the circuit switched on for?

.... s

6 A circuit is switched on for 15 minutes. During this time, 1200 C of electric charge are transferred.

Calculate the current flowing through the circuit.

.... A

Current, resistance and potential difference

Current, resistance and potential difference

The current in an electrical circuit depends on:
- the resistance in the circuit
- the potential difference in the circuit.

For a given potential difference across a component, the greater the resistance of the component, the smaller the electric current.

Calculating potential difference

Current, resistance and potential difference are related by the following equation:

$$V = IR$$

where

V is the potential difference, in volts (V)

I is the electric current, in amperes (or amps) (A)

R is the resistance, in ohms (Ω)

Remember that the current is measured in series with the component using an ammeter, and the potential difference is measured in parallel using a voltmeter.

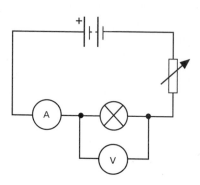

A current of 10 A flows through a circuit with a resistance of 20 Ω.

Calculate the potential difference of the circuit.

Begin by listing the information known about the circuit.

Current = 10 A

Resistance = 20 Ω

Substitute these values into the equation for potential difference.

$V = IR = 10 \times 20$

$V = 200\,V$

Rearranging the equation

Sometimes the current or the resistance will not be known, so you will need to rearrange the equation $V = IR$ to make the unknown quantity the subject. Remember that when rearranging equations, whatever is done to one side must be done to the other.

If the current is unknown	$V = IR$ $\frac{V}{R} = I$	Divide both sides of the equation by resistance.
If the resistance is unknown	$V = IR$ $\frac{V}{I} = R$	Divide both sides of the equation by current.

A circuit with a current of 0.5 A flowing through it displays a reading of 4 V on the voltmeter. Calculate the resistance of the circuit.

Begin by listing the information known about the circuit.

Current = 0.5 A

Potential difference = 4 V

Substitute these values into the equation for electric current.

$V = IR$

$R = \frac{V}{I} = \frac{4}{0.5}$

$R = 8\,\Omega$

The electrical current, potential difference and resistance are related by the equation $V = IR$

Current, resistance and potential difference

Current, resistance and potential difference

1 For a constant potential difference, state the effect of the resistance decreasing on the value of the electric current.

....

Calculating potential difference

2 In words, write the equation that links current, potential difference and resistance.

....

3 The total resistance in a circuit is 300 Ω and the current flowing is 0.005 A.

Calculate the potential difference of the circuit.

.... V

4 The current flowing in a circuit containing three resistors is 3.0 A.
The resistors have a combined resistance of 250 Ω.

Calculate the potential difference of the circuit.

.... V

Rearranging the equation

5 The reading on a voltmeter in a circuit is 12 V and the circuit has a total resistance of 4 Ω.

Calculate the current flowing through the circuit.

.... A

6 A circuit with a current of 10 A flowing through it displays a reading of 400 V on the voltmeter.

Calculate the resistance of the circuit.

.... Ω

2 Resistors

Resistance

Resistance is a measure of how hard it is to get a current through a component at a particular potential difference.

Current–potential difference graphs show how the current through the component varies with the potential difference across it.

Resistors

For **ohmic conductors**, at a constant temperature, current is directly proportional to potential difference. This means that the value for resistance remains constant as the current changes. This is regardless of the direction that the current is flowing in.

A piece of copper wire is an example of an ohmic conductor. Resistors, such as the one shown in the circuit opposite, are also ohmic conductors.

For **non-ohmic conductors**, the resistance changes when the current changes. LDRs, thermistors, filament lamps and diodes are all examples of non-ohmic conductors.

Notice how the line on the graph is a **straight line** that **passes through the origin**. This tells us that the current is **directly proportional** to the potential difference.

For some resistors, known as ohmic conductors, the value for resistance remains constant. For non-ohmic conductors, the value for resistance can change as the current changes.

The resistance of a light dependent resistor (LDR) depends on the amount of light falling on it. Its resistance decreases as the amount of light falling on it increases. This allows more current to flow. This can be useful for switching on lights automatically when it gets dark.	The resistance of a thermistor depends on its temperature. Its resistance decreases as its temperature increases. This allows more current to flow. This can be useful in thermostats, which are used to control the heating in homes.

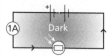 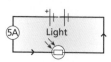 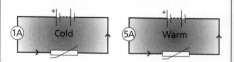

As the temperature of the filament lamp increases, and the bulb gets brighter, then the resistance of the lamp increases. This is regardless of which direction the current is flowing.	A diode allows a current to flow through it in one direction only. It has a very high resistance in the reverse direction so no current flows.

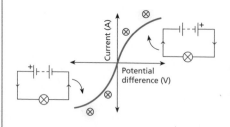 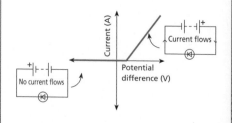

② Resistors

Resistance

1 What is meant by the term 'resistance'?

Resistors

2 How can you tell from a graph whether the values plotted against the x-axis and y-axis are **directly proportional** to one another?

3 Give an example of:

a) an ohmic conductor

b) a non-ohmic conductor

4 Sketch the shape of a current–potential difference graph for a filament lamp.

5 **a)** Sketch the shape of a current–potential difference graph for a diode.

b) Explain the shape of the graph in terms of resistance.

6 Suggest a use for:

a) a thermistor

b) an LDR

2 Series and parallel circuits

Series circuits

In a **series circuit**, all components are connected one after the other in one loop, going from one terminal of the battery to the other.

The same **current** flows through each component, i.e. $A_1 = A_2 = A_3$

In this circuit, both bulbs have the same **resistance** so the voltage is divided equally. If one bulb had twice the resistance of the other, then the voltage would be divided differently, e.g. 2 V and 1 V

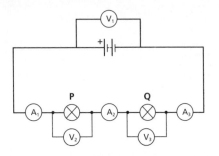

The **potential difference** supplied by the battery is divided up between the components in the circuit, i.e. $V_1 = V_2 + V_3$

The **total resistance** is the sum of the individual resistances of the components, i.e. P + Q. If both P and Q each have a resistance of 15 Ω, the total resistance would be 15 Ω + 15 Ω = 30 Ω

Parallel circuits

In a **parallel circuit**, all components are connected separately in their own loop going from one terminal of the battery to the other.

The total **current** in the main circuit is equal to the sum of the currents through the separate components, i.e. $A_1 = A_2 + A_3 = A_4$.

The **potential difference** across each component is the same (and is equal to the p.d. of the battery), i.e. $V_1 = V_2 = V_3$.

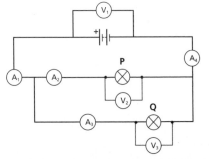

The **total resistance** of the circuit is less than the resistance of the smallest individual resistor. This is because, in a parallel circuit, the current has more paths through which it can flow.

Comparing series and parallel circuits

	Series circuit	Parallel circuit
Current	The current is the **same** through each component.	Total current is the **sum** of the currents through each component.
Potential difference	Total potential difference from the power supply is **shared** between components.	The potential difference is the **same** across each component.
Resistance	The total resistance is the **sum** of the individual resistances of each component.	The total resistance is **less than** the resistance of the smallest individual resistor.

In series circuits, all components are connected in one loop. In parallel circuits, the components are connected separately in their own loop.

2 Series and parallel circuits

Series circuits

1 A series circuit is shown. Write down the missing current reading for the second ammeter.

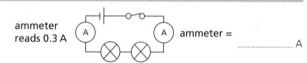

ammeter reads 0.3 A

ammeter = ... A

2 Two identical bulbs are connected to a 4.5 V battery in series.

Assuming that the bulbs have the same resistance, what will be the potential difference across each bulb in the circuit? Show clearly how you worked out the answer.

.... V

Parallel circuits

3 How are the components in a parallel circuit connected?

....

4 In the circuit shown, two lamps are connected to a cell. The ammeters X, Y and Z measure the current flowing at different points in the circuit.

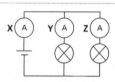

Which ammeter, X, Y or Z, will have the highest reading? Explain your answer.

....

....

....

Comparing series and parallel circuits

5 Complete the table comparing series and parallel circuits by writing the missing words in the spaces.

	Series circuit	Parallel circuit
Current	The current is the .... through each component.	Total current is the .... of the currents through each component.
Potential difference	Total potential difference from the power supply is .... between components.	The potential difference is the .... across each component.
Resistance	The total resistance is the .... of the individual resistances of each component.	The total resistance is .... than the resistance of the smallest individual resistor.

Direct and alternating potential difference and mains electricity

Direct and alternating potential difference

Direct potential difference produces a constant electric current that flows in one direction.

If current or voltage are plotted against time, the graph shows a constant horizontal line.

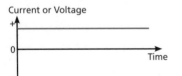

Alternating potential difference produces an alternating current that regularly changes direction.

If current or voltage are plotted against time for alternating potential difference, you can see the regular change in direction.

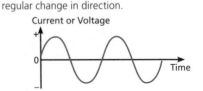

Direct potential difference produces a direct current.
Alternating potential difference produces an alternating current.

Mains electricity

Mains electricity is transmitted by an alternating potential difference supply. This is because the electricity can be transmitted more efficiently this way.

The frequency is the number of complete cycles of reversal per second. The UK mains electricity is 50 cycles per second (hertz). The UK mains supply has a voltage of about 230 volts. This voltage, if it isn't used safely, can kill.

The three-core cable

Most appliances connect to the mains electricity supply using a cable and a three-pin plug. The plug inserts into a socket on the ring main circuit.

The materials used for the plug and cable depend on their properties:

- The inner cores of the wires are made of copper because it's a good conductor.
- The outer layers are made of flexible plastic because it's a good insulator.
- The pins of a plug are made from brass because it's a good conductor.
- The casing is made from plastic or rubber because both are good insulators.

You might be asked to state **where** each wire in the three-core cable should be connected inside a plug.

Remember that:

- the **bl**ue wire is connected to the **b**ottom **l**eft
- the **br**own wire is connected to the **b**ottom **r**ight
- this leaves the green and yellow wire connected in the centre.

Earth wire (green and yellow) ———

Neutral wire (blue) – completes the circuit to allow current to flow to the device

Cable grip – secures cable in the plug

Casing ←

Fuse

Live wire (brown) – completes the circuit to allow current to flow to the device

Cable ←

Wire	Function
Live	Carries alternating current from the supply to the device
Neutral	Completes the circuit to allow current to flow to the device
Earth	Safety wire; this stops the appliance from becoming live

The potential difference between the live wire and the earth wire is 230V. The neutral wire has the same potential difference as the earth wire: 0V. The earth wire only carries an electric current if there is a fault in the appliance. This prevents the appliance from becoming live and therefore very dangerous to touch.

Direct and alternating potential difference and mains electricity

Direct and alternating potential difference

1 What is meant by an 'alternating potential difference'?

....

....

2 On the axes below, sketch the voltage–time graph for direct potential difference.

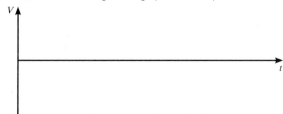

Mains electricity

3 Write down the frequency and the potential difference of the mains electricity supply.

Frequency: ....

Potential difference: ....

4 Is the mains electricity supply given by an alternating or a direct potential difference? Give a reason for your answer.

....

....

....

The three-core cable

5 The inner cores of the cables in a three-core cable are made from copper.

Why is copper used?

....

....

6 Complete the table below showing the functions of each wire in a three-core cable.

Wire	Function
Live	Carries .... current from the .... to the ....
Neutral	.... the circuit to allow .... to flow to the device.
Earth	.... wire; this stops the appliance from becoming ....

② Electrical power

What is electrical power?

The **power** of an electrical device is a measure of the rate of energy transfer by the device. For example, a microwave oven with a power rating of 900W transfers 900J of energy every second that it is turned on.

The greater the power of the device, the faster the rate of energy transfer.

Calculating electrical power

The power of a device depends on the potential difference supplied to it and the current flowing through it. The current, potential difference and power are related by the following equation:

$$P = IV$$

where

P is the power, in watts (W)

I is the current, in amperes (or amps) (A)

V is the potential difference, in volts (V)

Sometimes, the potential difference is not known but the resistance of the device is given instead.

As $V = I \times R$, it can be substituted into the equation for power to give $P = I \times I \times R$.

This gives an alternative equation for calculating power:

$$P = I^2 R$$

where

P is the power, in watts (W)

I is the current, in amperes (or amps) (A)

R is the resistance, in ohms (Ω)

a) An electric kettle is connected to the mains supply, with a potential difference of 230V. The current flowing through the kettle is 9A. Calculate the power of the kettle.

Begin by listing the information known about the kettle.

Potential difference = 230V
Current = 9A

Substitute these values into the equation for power.

$P = IV = 9 \times 230$
$P = 2070\,W$

b) An electric hairdryer is connected to the mains supply and has a resistance of 100Ω. The current flowing through the hairdryer is 5A. Calculate the power of the hairdryer.

List the information known about the hairdryer.

Resistance = 100Ω
Current = 5A

Substitute these values into the equation for power.

$P = I^2 R = 5^2 \times 100$
$P = 2500\,W$

Rearranging the equations

Sometimes you will need to rearrange the equations $P = IV$ or $P = I^2R$ to make the unknown quantity the subject

If the current is unknown		If the potential difference is unknown		If the resistance is unknown	
$P = IV$ $I = \dfrac{P}{V}$	Divide both sides of the equation by V.	$P = IV$ $V = \dfrac{P}{I}$	Divide both sides of the equation by I.	$P = I^2R$ $\dfrac{P}{I^2} = R$	Divide both sides of the equation by I^2.
Or					
$P = I^2R$ $\dfrac{P}{R} = I^2$	Divide both sides of the equation by R.				
$\sqrt{\dfrac{P}{R}} = I$	Take the square root of both sides.				

Electrical power can be calculated using the equations $P = IV$ and $P = I^2R$

② Electrical power

What is electrical power?

1 What is meant by 'electrical power'?

....

....

Calculating electrical power

2 A television is connected to the mains electricity supply, with a potential difference of 230 V. When it is switched on, the current flowing through the television is 0.7 A.

Calculate the power of the television.

.... W

3 A vacuum cleaner has a resistance of 35 Ω and a current of 8.45 A flowing through it.

Calculate the power of the vacuum cleaner.

.... W

Rearranging the equations

4 An electric motor has a potential difference of 250 V and a power rating of 8 kW.

Calculate the current flowing through the motor when it is switched on.

.... A

5 A bulb is connected in a circuit. The current flowing through the bulb is 3 A and the lamp has a power rating of 4.5 W.

Calculate the potential difference of the power supply in the circuit.

.... V

6 A vacuum cleaner has a power rating of 2600 W and a current of 13 A.

Calculate the resistance of the vacuum cleaner. Give your answer to 3 significant figures.

.... Ω

7 A microwave oven has a power rating of 900 W and a resistance of 100 Ω.

Calculate the current flowing through the microwave when it is switched on.

.... A

2) Energy transfers in appliances

Energy transfers in everyday appliances

Electrical devices are used to transfer energy. When an appliance is switched on, charge flows around the circuit. When charge flows around an electric circuit, work is done and so energy is transferred.

The amount of electrical energy transferred by an appliance depends on **two** factors:
- how long the appliance is switched on for
- the power of the appliance.

Calculating energy transfer

The amount of electrical energy transferred by an appliance depends on the power and the time it is switched on for. The energy transferred, power and time are related by the following equation:

$$E = Pt$$

where

E is the energy transferred, in joules (J)

P is the power, in watts (W)

t is the time, in seconds (s)

An electric kettle with a power rating of 2200 W takes $2\frac{1}{2}$ minutes to boil when it is filled with water. Calculate the energy transferred by the kettle in this time.

Begin by listing the information known about the kettle.

Power = 2200 W
Time taken = $2\frac{1}{2}$ minutes = 150 seconds

Substitute these values into the equation for energy transferred.

$E = Pt$
$E = 2200 \times 150$
$E = 330\,000\,J$

Sometimes the power or the time will not be known, so you will need to rearrange the equation $E = Pt$ to make the unknown quantity the subject. Remember that when rearranging equations, whatever is done to one side must be done to the other.

If the power is unknown	If the time is unknown
$E = Pt$ Divide both sides of the equation by time. $\frac{E}{t} = P$	$E = Pt$ Divide both sides of the equation by power. $\frac{E}{P} = t$

You know that $E = Pt$

⬇

Since $P = IV$, you can substitute this into the equation to give:
$E = VIt$

⬇

Since $Q = It$, you can substitute this into the equation to give:
$E = VQ$

where E = energy transferred, measured in joules (J)

V = potential difference, measured in volts (V)

Q = electric charge, measured in coulombs (C)

Energy transferred can be calculated using the equations $E = Pt$ and $E = VQ$

An electric oven, connected to the mains supply at 230 V, transfers 93 600 C of charge over a period of 2 hours. Calculate the energy transferred by the oven in this time. Give your answer in kilojoules (kJ).

List the information known about the oven.

Potential difference = 230 V
Charge = 93 600 C

Substitute these values into the equation for energy transferred.

$E = VQ$
$E = 230 \times 93\,600$
$E = 21\,528\,000\,J$
$E = 21\,528\,kJ$

② Energy transfers in appliances

Energy transfers in everyday appliances

1 State the **two** factors that the amount of energy transferred by an electrical appliance depends on.

....

....

Calculating energy transfer

2 In words, write the equation that links energy transferred, power and time.

....

3 A pair of hair straighteners, with a power rating of 60W, is switched on for 20 minutes.

Calculate the energy transferred by the hair straighteners in this time.

.... J

4 A 1500W food blender transfers 180kJ of energy when it is used.

Calculate the time the blender is used for. Give your answer in minutes.

.... minutes

5 A laptop transfers 702kJ of energy when it is used for a period of 3 hours.

Calculate the power of the laptop.

.... W

6 In words, write the equation that links charge, energy transferred and potential difference.

....

7 A desk lamp, connected to the mains electricity supply at 230V, transfers 6000C of electric charge.

Calculate the energy transferred by the lamp.

.... J

8 An electric iron is connected to the mains supply (230V) and transfers 165.6kJ of energy in the time it is used.

Calculate the charge transferred by the electric iron.

.... C

9 A bulb is connected to a circuit and, when it is switched on, 150J of energy are transferred. The charge transfer is 100C.

Calculate the voltage of the power supply for the circuit.

.... V

② The National Grid

What is the National Grid?

The National Grid links power stations to consumers via a system of **transformers** and **cables**. It is only the transformers and cables that make up the National Grid, **not** the power stations and homes.

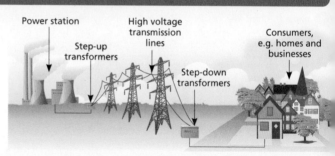

Power station | High voltage transmission lines | Consumers, e.g. homes and businesses
Step-up transformers
Step-down transformers

Parts of the National Grid

Electricity is generated at a potential difference of around 25 000 V in a power station outside of the National Grid.

Step-up transformers are used to increase the potential difference to around 400 000 V.

Increasing the potential difference to such high values means that the current through the cables remains very low. This significantly reduces energy loss through heating of the cables, meaning that the transmission is very **efficient**.

Step-up transformer

Iron core
Primary coil (input)
Secondary coil (output)

Since homes, offices, hospitals and other buildings are usually a long way from power stations, overhead **transmission cables** are used to transmit the electricity from the transformers. Pylons are used to raise these cables high above ground level to keep people safe.

The cables used are very thick to further reduce energy loss through them being heated. This reduces the **resistance** in the cables.

As $P = I^2 R$, keeping the current and the resistance low ensures that the power is delivered to the consumer as opposed to being dissipated into the surroundings as heat. However, voltages of 400 000 V are incredibly dangerous and cannot be used safely in homes and offices.

Step-down transformer

Iron core
Primary coil (input)
Secondary coil (output)

Step-down transformers are used to decrease the potential difference down to 230 V; a safe level if used correctly.

Step-up transformers are used to increase the potential difference so that electricity is transmitted at a low current, reducing energy loss by heating.

Step-down transformers are used to decrease the potential difference so that electricity can be safely used by consumers.

You can also calculate the size of the potential difference in each coil:

$$V_p I_p = V_s I_s \text{ where}$$

V_p is the potential difference in the primary coil and V_s is the potential difference in the secondary coil, in volts (V)

I_p is the current in the primary coil and I_s is the current in the secondary coil, in amperes (A)

What is the National Grid?

1. Which of the following accurately describes the National Grid? Tick the correct answer.

   A   Power stations and transmission cables linked to consumers

   B   Transmission cables connected to pylons

   C   Cables and transformers linking power stations to consumers

   D   Power stations, cables and transformers

Parts of the National Grid

2. What is the function of a **step-up** transformer in the National Grid?

3. Why is electricity transmitted at a voltage of 400000V, as opposed to a safer voltage of 230V?

4. What is the function of a **step-down** transformer in the National Grid?

5. A transformer has 50 turns on the primary coil. A potential difference of 230V is applied to the primary coil and a potential difference of 1000V is produced on the secondary coil.

   a) How many turns are there on the secondary coil?

   .... turns

   b) Is it a **step-up** or a **step-down** transformer? ....

6. The potential difference in the primary coil of a step-down transformer is 400000V. The potential difference in the secondary coil is 230V and the current in this coil is 13A.

   Calculate the current in the primary coil. Give your answer in standard form.

   .... A

3 Density

Density and states of matter

Density is a measure of how heavy something is for its size (volume).

In dense materials:
- the particles are close together
- there is very little space between the particles.

Usually, for the same material:
- a gas is less dense than a liquid
- a liquid is less dense than a solid.

Density and states of matter

Gas   Liquid   Solid

Low density   High density

Water is an exception. Liquid water is more dense than solid water (ice). This is why ice floats on water.

Calculating density

The density of a material depends on:
- the mass of the material
- the volume taken up by the material.

Density can be calculated using the equation below:

$$\rho = \frac{m}{v}$$

where

ρ is the density, in kilograms per cubic metre (kg/m³)

m is the mass, in kilograms (kg)

v is the volume, in cubic metres (m³)

A concrete block has a mass of 4800 kg and a volume of 2 m³. Calculate the density of the concrete block.

Begin by listing what is known about the block.

Mass = 4800 kg   Volume = 2 m³

Substitute these values into the equation for density.

$\rho = \frac{m}{v} = \frac{4800}{2} = 2400$ kg/m³

To calculate density, you need to:
- measure mass – use a top pan balance to get a value of the mass of a substance
- measure volume.

If the shape is **irregular**, use a Eureka can and a measuring cylinder to measure the volume of the displaced water.

Sometimes you will need to rearrange the equation $\rho = \frac{m}{v}$ to make the unknown quantity the subject.

If the mass is unknown	$\rho = \frac{m}{v}$ $\rho \times v = m$	Multiply both sides of the equation by volume.
If the volume is unknown	$\rho = \frac{m}{v}$ $\rho \times v = m$	Multiply both sides of the equation by volume.
	$v = \frac{m}{\rho}$	Divide both sides of the equation by density.

Floating and sinking

When substances are heated, they expand and their density decreases. So, as air gets heated, the spaces between the gas particles increase and the air expands. This means there are the same number of particles but they take up a greater volume, and so density decreases. The warm air rises. This is how a hot air balloon rises.

A less dense material floats on top of a more dense material.

Cooking oil has a density of 0.93 g/cm³ and water has a density of 1.00 g/cm³, so the oil floats on water.	Sand has a density of 1.5 g/cm³ and so it sinks in water.

Density

Density and states of matter

1 What **two** factors does the density of an object depend on?

2 Using particle diagrams, explain why the density of a solid substance tends to be greater than the density of the same mass of the same substance in the gas state.

Calculating density

3 In words, write the equation that links density, mass and volume. Give the units that each quantity is measured in.

4 A lead block with a volume of 0.1 m³ has a mass of 1134.3 kg.

Calculate the density of the block.

.... kg/m³

5 A gold ring has a mass of 7.5 g. The density of gold is 19 320 kg/m³.

Calculate the volume of gold used to make the ring.

.... m³

Floating and sinking

6 An iron nail is dropped into a beaker of water. The density of water is 998 kg/m³ and the density of iron is 7860 kg/m³.

Use this information to decide whether the iron nail will float or sink in water. Explain your answer.

Changes of state and internal energy

Changing state

When a substance changes state, only the movement and position of the particles change. This means that changing state is a physical change.

Substances can change state if you add or remove heat (thermal energy).

Energy transferred to the particles from the surroundings by heat

Sublimation

Solid — Melting → Liquid — Boiling → Gas

Freezing ← Condensation ←

Energy transferred from the particles to the surroundings by heat

Melting and boiling points

The graph shows the heating of an ice cube.

The temperature increases as you heat the ice until the melting point is reached at 0°C.

The melting point is where a pure substance changes state from a solid to a liquid. At this point, the temperature does not change as all the heat (thermal energy) is being used to overcome the strong intermolecular forces between water particles in the ice.

> The melting and boiling points of an unknown substance can be compared to information on databases in order to identify it.

Once the ice has melted, the temperature rises again until the boiling point is reached at 100°C.

The boiling point is where a pure substance changes state from a liquid to a gas. At this point, the temperature does not change as all the energy is being used to pull the water particles apart to form steam.

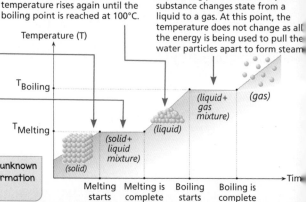

Temperature (T)

$T_{Boiling}$

$T_{Melting}$

(solid)

(solid+ liquid mixture)

(liquid)

(liquid+ gas mixture)

(gas)

Time

Melting starts | Melting is complete | Boiling starts | Boiling is complete

Evaporation and boiling

Both evaporation and boiling involve changing a liquid into a gas, but they have differences.

Evaporation	Boiling
• Occurs at any temperature between the melting point and boiling point of the substance • Is a slow process • Only happens at the surface • Energy comes mainly from kinetic energy of the substance	• Only happens at boiling point • Whole liquid changes to a gas • Energy comes mainly from the surroundings of the substance

Internal energy

Internal energy is equal to the sum of all of the kinetic energy and potential energy of the particles within a system.

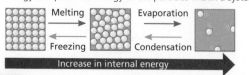

Melting → Evaporation →

← Freezing ← Condensation

Increase in internal energy

When a substance is heated up, the particles in the system gain energy. This change in energy within the system either

• causes the temperature of the substance to rise, or
• changes the state of the substance.

Changes of state and internal energy

Changing state

1 Complete the diagram by adding the names of the state changes in the answer spaces.

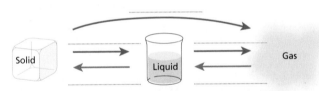

Melting and boiling points

2 **a)** Define the term 'melting point'.

....

....

b) Define the term 'boiling point'.

....

....

3 Sketch a temperature–time graph for a substance being heated and changing from a solid to a liquid and then finally to a gas.

Label the graph with the melting point and boiling point of the substance.

Evaporation and boiling

4 State **one similarity** and **one difference** between evaporation and boiling.

Similarity: ....

....

Difference: ....

....

....

Internal energy

5 Define the term 'internal energy'.

....

....

Temperature changes in a system and specific heat capacity

Temperature changes in a system

When the temperature of a substance increases, the extent to which it increases depends on:

- the mass of the substance heated
- the type of substance being heated
- the energy input to the system.

If the same energy were inputted to the water in the beaker and the water in the bath tub, the temperature rise of the water in the beaker would be greater than that in the bath tub, despite water being the substance heated in both cases. This is because the mass of the water in the beaker is significantly less than the mass of the water in the bath tub.

Specific heat capacity

Specific heat capacity is the amount of energy required to raise the temperature of 1 kg of a substance by 1°C. This means that, with the same energy input, some substances will experience a greater increase in temperature than others.

The higher the specific heat capacity, the smaller the temperature rise for the same energy input.

Examples of specific heat capacities

Copper	389 J/kg°C
Aluminium	900 J/kg°C
Water	4200 J/kg°C

Calculations involving temperature changes in a system

The input energy, mass, specific heat capacity and temperature change are linked by the equation below:

$$\Delta E = mc\Delta\theta$$

where

ΔE is the change in thermal energy, in joules (J)

m is the mass, in kilograms (kg)

c is the specific heat capacity, in joules per kilogram per degree Celsius (J/kg°C)

$\Delta\theta$ is the change in temperature, in degrees Celsius (°C)

Sometimes you will need to rearrange the equation $\Delta E = mc\Delta\theta$ to make the unknown quantity the subject.

If the mass is unknown	$\Delta E = mc\Delta\theta$ $\dfrac{\Delta E}{(c\Delta\theta)} = m$	Divide both sides of the equation by $(c\Delta\theta)$
If the change in temperature is unknown	$\Delta E = mc\Delta\theta$ $\dfrac{\Delta E}{(mc)} = \Delta\theta$	Divide both sides of the equation by (mc)
If the specific heat capacity is unknown	$\Delta E = mc\Delta\theta$ $\dfrac{\Delta E}{(m\Delta\theta)} = c$	Divide both sides of the equation by $(m\Delta\theta)$

a) A copper coin has a mass of 5.5 g and, when heated, its temperature rises from 15°C to 80°C. The specific heat capacity of copper is 389 J/kg°C. Calculate the change in thermal energy of the coin.

Begin by listing the information known about the coin

Mass = $\dfrac{5.5}{1000}$ = 5.5 × 10^{-3} kg

Specific heat capacity = 389 J/kg°C

Change in temperature = 80°C − 15°C = 65°C

Substitute these values into the equation

$\Delta E = mc\Delta\theta$

ΔE = (5.5 × 10^{-3}) × 389 × 65

ΔE = 139.1 J

b) 315 kJ of energy are supplied to a 1.5 kg brick. Brick has a specific heat capacity of 840 J/kg°C. Calculate the change in temperature of the brick.

Begin by listing the information known about the brick

Mass = 1.5 kg

Specific heat capacity = 840 J/kg°C

Change in thermal energy = 315 kJ = 315 000 J

Substitute these values into the equation

$\Delta E = mc\Delta\theta$

$\dfrac{\Delta E}{(mc)} = \Delta\theta$

$\Delta\theta = \dfrac{315\,000}{(1.5 \times 840)}$

$\Delta\theta$ = 250°C

Temperature changes in a system and specific heat capacity

Temperature changes in a system

1 When a substance is heated with no change of state, its temperature will rise.

State the **three** factors that this size of temperature rise will depend on.

_____

_____

_____

Specific heat capacity

2 Define the term 'specific heat capacity'.

_____

_____

3 Assuming that two substances have an equal mass, will the substance with a higher or lower specific heat capacity have a greater temperature rise for the same input energy?

_____

Calculations involving temperature changes in a system

4 In words, write the equation that links the change in thermal energy, change in temperature, mass and specific heat capacity.

_____

5 Cooking oil has a specific heat capacity of 2000 J/kg°C.

How much energy would need to be supplied to 5 g of cooking oil to raise its temperature from 20°C to 100°C?

_____ J

6 146.25 kJ of energy are supplied to 250 g of engine oil. The temperature of the oil rises from 20°C to 50°C.

Calculate the specific heat capacity of the engine oil in J/kg°C.

_____ J/kg°C

7 101.5 kJ of energy are supplied to an aluminium saucepan. As a result, the temperature of the saucepan increases from 15°C to 220°C. The specific heat capacity of aluminium is 900 J/kg°C.

Calculate the mass of the saucepan in grams.

_____ g

Changes of state and specific latent heat

Changes of state

The energy needed to cause a substance to change state is known as the **latent heat**. Latent heat is the energy that is added to a substance as it changes state; it is the energy needed to overcome the intermolecular forces.

The energy supplied during a state change increases the energy stored (the internal energy) but not the temperature.

Specific latent heat

The specific latent heat is the energy required to change the state of 1 kg of a substance without a change in temperature.

Just as different substances have different values for specific heat capacity, they also have different values for specific latent heat.

The **specific latent heat of fusion** is the energy required to melt 1 kg of a substance without a change in temperature.

The **specific latent heat of vaporisation** is the energy required to turn 1 kg of a liquid substance into a gas, without a change in temperature.

Calculations involving specific latent heat

The energy required to change the state of a substance depends on:

- the mass of the substance
- the specific latent heat of the substance.

The energy, mass and specific latent heat are related by the following equation:

$$E = mL$$

where

E is the energy supplied, in joules (J)

m is the mass of the substance, in kilograms (kg)

L is the specific latent heat of the substance, in joules per kilogram (J/kg)

Sometimes you will need to rearrange the equation $E = mL$ to make the unknown quantity the subject.

If the mass is unknown		If the specific latent heat is unknown	
$E = mL$ $m = \dfrac{E}{L}$	Divide both sides of the equation by L.	$E = mL$ $L = \dfrac{E}{m}$	Divide both sides of the equation by m.

> The energy, mass and specific latent heat are linked by the equation $E = mL$

a) An ice cube has a mass of 7 g. The specific latent heat of fusion of water is 334 000 J/kg. Calculate the energy required to completely melt the ice cube.

List the information known about the ice cube.

Mass = $\dfrac{7}{1000}$ = 7 × 10⁻³ kg

Specific latent heat = 334 000 J/kg

Substitute these values into the equation for specific latent heat.

$E = mL = (7 \times 10^{-3}) \times 334\,000$
$E = 2338\,J$

b) A kettle has a capacity of 1.5 litres. When full, there is 1.5 kg of water in the kettle. Calculate the energy required to boil all of the water at 100°C in the kettle, given that the specific latent heat of vaporisation of water is 2 260 000 J/kg.

Begin by listing the information known about the water.

Mass = 1.5 kg

Specific latent heat = 2 260 000 J/kg

Substitute these values into the equation.

$E = mL = 1.5 \times 2\,260\,000$
$E = 3\,390\,000\,J = 3390\,kJ$

c) It takes 66 J of energy to melt a copper coin. The specific latent heat of fusion of copper is 13 200 J/kg. Calculate the coin's mass.

List the information known about the coin.

Energy = 66 J

Specific latent heat = 13 200 J/kg

Substitute these values into the equation.

$m = \dfrac{E}{L} = \dfrac{66}{13\,200}$
$m = 5 \times 10^{-3}$ kg

Changes of state and specific latent heat

Changes of state

1 Define the term 'latent heat'.

....

....

....

Specific latent heat

2 The specific latent heat of fusion and the specific latent heat of vaporisation can be used to calculate the energy required to change the state of a substance.

What is the difference between the specific latent heat of fusion and the specific latent heat of vaporisation?

....

....

....

Calculations involving specific latent heat

3 In words, write the equation that links energy, mass and specific latent heat.

....

4 Milk can be frozen to reduce wastage. Milk has a specific latent heat of fusion of 275 700 J/kg. Calculate the energy required to freeze 20 g of milk, assuming no change in temperature.

.... J

5 An aluminium block has a mass of 5 kg.

If 1995 kJ of energy are supplied in order to melt the block, calculate the specific latent heat of fusion of aluminium.

.... J/kg

6 1710 kJ of energy are needed to boil a sample of liquid lead. The specific latent heat of vaporisation of lead is 855 000 J/kg.

Calculate the mass of the lead sample.

.... kg

Particle motion in gases

The particle model

Anything that takes up space and mass is called **matter**. The particle model can be used to represent how matter is arranged and moves in each state of matter.

Solid

Liquid

Gas

Like all models, the particle model has limitations:

- particles are identical unless they are given different colours
- the individual atoms in the particles are not shown
- intermolecular forces between particles are not shown.

Particle motion in gases

The circles in the diagram represent the gas particles in a balloon.

The **direction** of the arrows represents the **direction** of movement of the particles and the **length** of the arrows represents the **speed** of the particles.

In conclusion, the particles in a gas:

- are in constant motion, with the speed and direction of the particles being random
- move at a range of speeds.

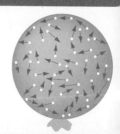

Motion, temperature and pressure in gases

The temperature of a gas is related to the **average kinetic energy** of all the gas particles in the substance. An increase in temperature leads to an increase in the average kinetic energy of the particles.

If the speed of the particles increases, so does the kinetic energy.

Changing the temperature of a gas, at constant volume, changes the pressure exerted by the gas. Pressure is caused by collisions between the gas particles and the walls of the container in which it is held.

If the temperature is increased, the average kinetic energy of the particles increases. This means that they will collide with the walls of the container **more frequently** and with **more energy**. The increasing frequency of collisions raises the pressure of the gas, since each collision creates a force.

You can calculate kinetic energy using the following equation:

$$E_k = \tfrac{1}{2}mv^2$$

where

E_k = kinetic energy, measured in joules (J)
m = mass, measured in kilograms (kg)
v = speed, measured in metres per second (m/s)

On average, the arrows are longer when the temperature is increased. This represents an increase in the speed of the particles.

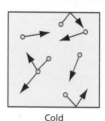

Cold

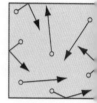

Hot

Particles in a gas move in constant, random motion and at a range of speeds. As the temperature of a gas held at constant volume is increased, the pressure of the gas also increases.

③ Particle motion in gases

The particle model

1 State **two** limitations of the particle model of matter.

Solid

Liquid

Gas

....

....

Particle motion in gases

2 Describe the particle motion in gases. In your answer, refer to the direction and speed of the particles.

....

....

....

Motion, temperature and pressure in gases

3 Explain the effect of increasing the temperature of a gas on the average kinetic energy of the particles in the gas.

....

....

4 Using the equation for kinetic energy, explain the effect of doubling the average speed of gas particles on the kinetic energy stored.

....

....

....

5 Explain what causes pressure when a gas is in a container.

....

....

6 Explain why, for a fixed mass of gas, an increase in temperature leads to an increase in pressure.

....

....

....

Structure of the atom, mass and atomic number, and isotopes

Structure of the atom

An atom is the smallest particle of an element that can exist. Atoms are tiny, with diameters of about 1×10^{-10} m.

An atom has **protons**, **neutrons** and **electrons**. All atoms of a particular element have the same number of protons.

Atomic particle	Relative mass	Relative charge
Proton	1	+1
Neutron	1	0
Electron	0 (nearly)	−1

An atom has the same number of protons as electrons. So, an atom as a whole has no electrical charge.

The electrons are located in energy levels at varying distances from the nucleus.

The nucleus of an atom contains the protons and neutrons. The radius of the nucleus is about 10 000 smaller than the radius of the atom.

The diameter of a gold nucleus is 3×10^{-14} m. The diameter of a sodium nucleus is 3×10^{-15} m. How many times larger is the gold nucleus than the sodium nucleus?

Ratio $= \dfrac{3 \times 10^{-14}}{3 \times 10^{-15}} = 10$

So, the gold nucleus is 10 times larger than the sodium nucleus.

Calculate the ratio of the diameters of the nuclei to compare one to the other.

The electron configuration can be altered by the absorption or emission of electromagnetic radiation:

Electron
Neutron
Proton

- If electromagnetic radiation is absorbed by the atom, an electron may move to a higher energy level further away from the nucleus.
- If electromagnetic radiation is emitted, an electron falls down to a lower energy level.

Mass and atomic number

Atoms of different elements have different numbers of protons.

The number of protons defines the element:
- The number of protons in an atom is called its **atomic number**.
- The number of protons and neutrons in an atom is called its **mass number**.

Mass number → 4
Atomic number → 2

$^{4}_{2}\text{He}$

Element symbol

To calculate the number of...	protons	electrons	neutrons
Use...	the atomic number	the atomic number	the mass number minus the atomic number

Calculate the number of neutrons in a sodium nucleus.

$^{23}_{11}\text{Na}$

To calculate the number of neutrons, subtract the atomic number from the mass number.

The mass number is 23 and the atomic number is 11.

The number of neutrons = 23 − 11 = 12

Isotopes

Some atoms of the same element can have different numbers of neutrons. These are called **isotopes**.

> Atoms have a nucleus that contains protons and neutrons, surrounded by electrons in energy levels. Isotopes are atoms of the same element with the same number of protons, but a different number of neutrons.

Calculate the number of neutrons in each of the three isotopes of carbon:

$^{12}_{6}\text{C}$ $^{13}_{6}\text{C}$ $^{14}_{6}\text{C}$

Subtract the atomic number from the mass number for each isotope.

Carbon-12: No. of neutrons = 12 − 6 = 6
Carbon-13: No. of neutrons = 13 − 6 = 7
Carbon-14: No. of neutrons = 14 − 6 = 8

Structure of the atom, mass and atomic number, and isotopes

Structure of the atom

1. Complete this table showing the relative masses and charges of the three subatomic particles.

Atomic particle	Relative mass	Relative charge
Proton		
Neutron		
Electron		

2. The diameter of a uranium nucleus is 1.5×10^{-14} m.
   The diameter of a sodium nucleus is 1.7×10^{-15} m.

   How many times bigger is a uranium nucleus than a helium nucleus?

   ....

Mass and atomic number

3. Find the number of neutrons in an aluminium nucleus, given this information:

   $^{27}_{13}$Al

   ....

4. What can be deduced about a magnesium atom from this information?

   $^{24}_{12}$Mg

   Number of: protons = ...   electrons = ...   neutrons = ...

Isotopes

5. What is meant by the term 'isotope'?

   ....

   ....

   ....

6. Give **one similarity** and **one difference** in structure of these isotopes of hydrogen.

   $^{1}_{1}$H

   hydrogen-1 (protium)

   $^{2}_{1}$H

   hydrogen-2 (deuterium)

   $^{3}_{1}$H

   hydrogen-3 (tritium)

   Similarity: ....

   ....

   Difference: ....

   ....

4 Development of the model of the atom

The solid sphere model

Theories about what matter consists of have been developing since the time of Greek philosopher Democritus (circa 460–370 BCE).

Prior to the discovery of the electron, atoms were thought to be tiny, solid spheres that could not be divided further. The term **atom** comes from the Greek adjective, *atomos*, which means **indivisible**.

The plum pudding model

In 1897, JJ Thomson discovered the **electron**. The discovery of the electron led to the model of the atom being revised.

He proposed that atoms consisted of a positively charged sphere, in which negatively charged electrons were scattered.

This model became known as the '**plum pudding model**' as it was thought that the electrons represented plums embedded into a sponge pudding.

The nuclear model

In 1911, Ernest Rutherford carried out some experiments in which alpha particles were directed at a thin film of gold foil.

This led to the suggestion that:
- most of the atom was empty space, as most alpha particles passed straight through the foil
- the centre of the atom contained a positively charged nucleus, as some of the positively charged alpha particles were deflected
- the nucleus was extremely tiny, and that this was where all the mass and charge of the atom was contained.

This became known as the **nuclear model** of the atom.

Atoms

A very small number of alpha particles deflected straight back

Most alpha particles passed straight through

Some alpha particles deflected through at small angles

New experimental evidence can lead to a scientific model being changed or replaced.

The Bohr model

In 1913, Niels Bohr adapted the nuclear model of the atom to put electrons orbiting at specific distances from the nucleus. His calculations agreed with experimental observations at the time and so the model of the atom was again revised.

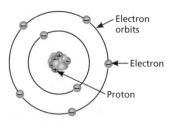

Electron orbits

Electron

Proton

Discovery of the neutron

In 1932, nearly 20 years after the nucleus became an accepted scientific idea, James Chadwick discovered the **neutron**. This led to the current model of the atom, which includes neutrons in the nucleus of an atom.

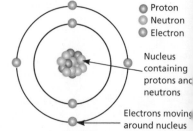

Proton
Neutron
Electron

Nucleus containing protons and neutrons

Electrons moving around nucleus

4 Development of the model of the atom

The solid sphere model

1 Describe the solid sphere model of the atom.

....

The plum pudding model

2 The discovery of which particle led to the plum pudding model being suggested? ....

3 Draw a diagram to represent the plum pudding model of the atom.

The nuclear model

4 Draw a line from each experimental observation on the left to the corresponding conclusion about the structure of the atom made from the alpha scattering experiment.

Experimental observation   **Conclusion**

Experimental observation	Conclusion
Most alpha particles passed straight through the gold foil	The nucleus is very small and very dense
Some alpha particles were deflected at small angles	Most of the atom is empty space
A very small number of alpha particles were deflected straight back	The nucleus is positively charged

The Bohr model

5 Draw a diagram to represent the Bohr model of the atom. Show the electrons and protons.

Discovery of the neutron

6 Who discovered the neutron in 1932? ....

Radioactive decay

Radioactive isotopes (radioisotopes or radionuclides) are atoms with unstable nuclei. They may disintegrate and emit **radiation**. This is **radioactive decay**.

Radioactive decay can result in the formation of a different atom with a different number of protons.

Two examples of radioactive decay are alpha (α) radiation and beta (β) radiation. A third type is gamma (γ) radiation. However, unlike alpha and beta, gamma emissions have no effect on the structure of the nucleus. Some stable nuclei may also emit a neutron.

Radioactive decay is a **random** process: it is impossible to tell **which** nuclei will decay or **when** a particular nucleus will decay.

> There are three types of nuclear radiation: alpha decay, beta decay and gamma emission.

Alpha (α) decay	Beta (β) decay
In alpha decay, the original atom decays by ejecting an alpha particle from the nucleus.	In beta decay, the original atom decays by changing a neutron into a proton and an electron.
An alpha particle is a helium nucleus – a particle made up of two protons and two neutrons.	The newly formed high-energy electron ejected from the nucleus is a β particle.
A new atom is formed with α decay.	A new atom is formed with β decay.

Unstable nucleus	New nucleus	α particle	Unstable nucleus	New nucleus	β particle

Ionisation

Radioactive particles can collide with neutral atoms or molecules. Electrons will be knocked out of their structure and they will become charged. These charged particles are called **ions**.

Alpha radiation and beta radiation are known as **ionising radiation**.

Ionising radiation can damage molecules in a healthy cell, causing the death of the cell.

Activity and count rate

The rate at which a source of unstable nuclei decay is known as the **activity** of the source. The **activity** is measured in **becquerels** (Bq).

The **count rate** is the number of decays per second that are recorded by a detector, such as a Geiger-Müller tube. Count rate is also measured in **becquerels**.

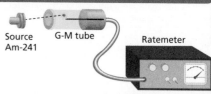

Source Am-241   G-M tube   Ratemeter

Properties of alpha, beta and gamma radiation

Type of radiation	Alpha	Beta	Gamma
Penetration power	Stopped by skin or a single sheet of paper	Stopped by 2–3 mm of aluminium foil	Stopped by thick lead or concrete
Range in air	< 5 cm	~ 1 m	> 1 km
Ionising power	High	Low	Very low

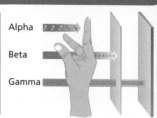

Alpha

Beta

Gamma

Skin Aluminium Lea

4 Radioactive decay and nuclear radiation

Radioactive decay

1 What is meant by the term 'radioactive decay'?

....

....

2 What is meant by the 'random nature' of radioactive decay?

....

....

3 What does an alpha particle consist of?

....

4 Explain why a new element is formed in the process of alpha decay.

....

....

....

....

5 In beta decay, a neutron decays inside an unstable nucleus.

Into which **two** particles does the neutron decay?

....

Ionisation

6 What effect can ionisation have on healthy cells in the body?

....

Activity and count rate

7 What is the difference between the 'activity' and the 'count rate' of a source?

....

....

Properties of alpha, beta and gamma radiation

8 Complete the table to compare the properties of alpha, beta and gamma radiation.

Type of radiation	Penetration power	Range in air	Ionising power
Alpha	Stopped by	< 5 cm	....
Beta	Stopped by 2–3 mm of aluminium foil	....	Low
Gamma	Stopped by		

4 Nuclear equations

What are nuclear equations?

Nuclear equations can be used to represent radioactive decay.

Loss of an alpha or a beta particle from the nucleus may result in a change in the mass of the nucleus and/or a change in the charge of the nucleus.

Gamma radiation is the emission of a high energy, high frequency electromagnetic wave which is uncharged and has no mass. Therefore, emission of a gamma wave from a nucleus does not cause the mass or the charge to change.

An alpha particle is represented by the symbol

$$^{4}_{2}\text{He}$$

A beta particle is represented by the symbol

$$^{0}_{-1}\text{e}$$

Remember that a beta particle is a fast-moving electron.

Nuclear equations to represent alpha and beta decay

Nuclear equations can be used to represent both alpha and beta decay.

A represents the mass number; Z represents the atomic number.

Alpha decay	Beta decay
The mass number of radon is 219. The atomic mass of polonium is 215. Alpha decay causes the **mass number to decrease by 4**.	The mass number of carbon is 14. The atomic mass of nitrogen is also 14. Beta decay causes the **mass number to remain unchanged**.

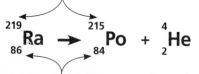

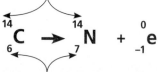

The atomic number for radon is 86, whereas the atomic number for the resultant nucleus, polonium, is 84. So alpha decay causes the **atomic number to decrease by 2**.

$$^{A}_{Z}X \rightarrow ^{A-4}_{Z-2}Y + ^{4}_{2}\text{He}$$

The atomic number for carbon is 6 and the atomic number for the resultant nucleus, nitrogen, is 7. So beta decay causes the **atomic number to increase by 1**.

$$^{A}_{Z}X \rightarrow ^{A}_{Z+1}Y + ^{0}_{-1}\text{e}$$

Balancing nuclear equations

Complete the nuclear equation for the decay of uranium-238 into thorium-234 by alpha decay adding the missing values, X and Y, into the equation shown.

$$^{238}_{92}U \rightarrow ^{234}_{Y}\text{Th} + ^{X}_{2}\text{He}$$

Begin by considering the atomic masses of the particles.

The atomic masses have to balance:
$238 = 234 + X$
$X = 238 - 234$
$X = 4$

Then consider the atomic numbers of the particles.

The atomic numbers have to balance, so:
$92 = Y + 2$
$Y = 92 - 2$
$Y = 90$

In nuclear equations, the atomic masses and atomic numbers must balance on either side of the arrow.

Nuclear equations

What are nuclear equations?

1 Write the symbols used to represent:

   a)   an alpha particle   ....

   b)   a beta particle   ....

2 Explain why the emission of a gamma wave from a radioactive nucleus **does not** result in a change in mass or charge of the nucleus.

....

....

....

....

Nuclear equations to represent alpha and beta decay

3 Complete the table to show the effect of alpha decay on the values for atomic number and atomic mass number.

	Atomic number	Atomic mass number
After alpha decay	Decreases by	.... by 4

4 Complete the table to show the effect of beta decay on the values for atomic number and atomic mass number.

	Atomic number	Atomic mass number
After beta decay	Increases by	

Balancing nuclear equations

5 Calculate the values for X and Y in the nuclear equation below.

$$^{239}_{Y}\text{Pu} \rightarrow \, ^{X}_{92}\text{U} + \, ^{4}_{2}\text{He}$$

$X =$ ....

$Y =$ ....

6 Balance the nuclear equation below.

$$^{}_{6}\text{C} \rightarrow \, ^{11}_{}\text{B} + \, ^{}_{-1}\text{e}$$

Half-lives and the random nature of radioactive decay

Half-life

The decay of a radioactive isotope is **random**, meaning that it is not possible to tell:

- when a particular nucleus will decay
- which nuclei will decay at any given time.

Despite the random nature, all radioactive isotope decays follow a similar pattern. If a sample contains 1000 nuclei, then after a certain amount of time 500 will remain. Once the same amount of time has passed again, 250 nuclei will remain. The time taken for the number of nuclei of an isotope in a sample to halve is the **half-life**.

Different radioactive isotopes have different half-lives, ranging from less than 1 second to more than 1000 years. An alternative way to define half-life is the time it takes for the count rate (or activity) of the sample to fall to half of its initial value.

> The half-life of a sample is the time taken for half of the nuclei in a sample to decay, or the time taken for the activity of the sample to fall to half of its original value.

Using graphs to determine half-life

1. Locate the point on the y-axis where the value has fallen to half the original value $\left(\frac{A_0}{2}\right)$.
2. Draw a horizontal straight line from this value to meet the curve.
3. Draw a vertical straight line down from the point where it meets the curve to the value on the x-axis. This gives you the first half-life, $t_{\frac{1}{2}}$.
4. Repeat, finding where the value on the y-axis falls by half again (so a quarter of its original value, $\frac{A_0}{4}$).
5. The time from the first half-life to this one is the second half-life.
6. Repeat for a third time and take an average of the half-lives.

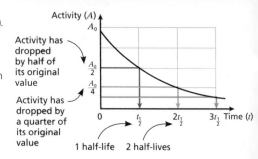

Activity has dropped by half of its original value

Activity has dropped by a quarter of its original value

1 half-life   2 half-lives

Using half-lives

a) A radioactive isotope has a half-life of 12 minutes. Assuming that there were 100 nuclei in the sample at the start, how many nuclei remain after 60 minutes?

Time (minutes)	Number of nuclei remaining
0	100
12 (after 1 half-life)	50
24 (after 2 half-lives)	25
36 (after 3 half-lives)	12
48 (after 4 half-lives)	6
60 (after 5 half-lives)	3

After 60 minutes, 3 nuclei remain.

b) The half-life of carbon-14 is 5700 years. If there are 10 g of carbon-14 in a sample, how much will remain after 22800 years?

$\frac{22800}{5700} = 4$ half-lives — Begin by calculating the number of half-lives that have passed.

$\left(\frac{1}{2}\right)^4 = \frac{1}{16}$
So, $\frac{1}{16}$ th will remain. — Now calculate the proportion of the sample that will remain after 4 half-lives.

Finally, calculate what $\frac{1}{16}$ th of 10 g is.

$\frac{1}{16} \times 10 = 0.625\,$g will remain after 22800 years.

The amount remaining of a sample can be expressed as a ratio, where the activity after x half-lives is compared to the initial activity:

Activity after x half-lives : initial activity

So, if $\frac{1}{25}$ th remains, this can be expressed as $1:25$

Half-lives and the random nature of radioactive decay

Half-life

1 State **two** ways in which radioactive decay can be described as 'random'.

_____

_____

2 What is the half-life of a radioisotope?

_____

_____

Using graphs to determine half-life

3 Determine the half-life of this isotope.

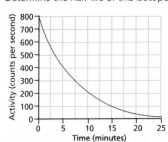

_____ minutes

Using half-lives

4 Cobalt-60 has a half-life of 5 days.

Assuming that a sample has an initial activity of 600 Bq, what will be the activity of the sample after 30 days?
Use a table (on a separate piece of paper if needed) in your answer.

_____ Bq

5 Caesium-137 has a half-life of 30 years.

a) What proportion of a 1 kg sample will remain after 120 years?

_____

b) How much of the sample will remain after 120 years?

_____ g

4 Radioactive contamination

What is meant by irradiation and contamination?

Irradiation		Contamination	
If an object is **irradiated**, it has been exposed to nuclear radiation: • the irradiated object does **not** become radioactive • the irradiation stops as soon as the source is removed.	Object exposed to radiation 	If an object is **contaminated**, there is a presence of material containing radioactive atoms on or in the object: • the contaminated object remains radioactive for as long as the source is on/in it • the hazard from contamination is due to the decay of the contaminating atoms.	Radiation placed in or on object

Irradiation

If an object is irradiated, living cells can be damaged. This can be both problematic and beneficial.

Sterilisation	Medical treatment
Gamma rays — Bacteria killed Fruit can be exposed to cobalt-60, which emits gamma rays. These gamma rays can kill bacteria on the surface of the fruit, preserving it and extending its shelf-life. In a similar way, irradiation is also used to sterilise surgical instruments.	Gamma rays can be directed towards the site of tumours to kill cancerous tissue. Care must be taken to calculate the correct dose as local healthy tissues can also be damaged in the process.  Gamma rays Protective head gear Cancerous tumour

Contamination

Contamination of an object can cause problems, but is useful in other ways.

Checking for leaks in underground pipes	Medical contamination
It is easy to see a leak in a water pipe above ground, but not if it is underground. If a leak is suspected, a gamma-emitting isotope can be introduced into the water supply. If water gathers at the site of a leak, the build-up of the gamma ray emission can be detected. This helps to identify the source of the leak. 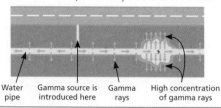 Water pipe — Gamma source is introduced here — Gamma rays — High concentration of gamma rays	Tracers, such as technicium-99, can be used for medical purposes. Medical tracers show up soft tissue as the gamma rays that are emitted by the tracer can pass through the body and be detected. If there is a blockage in a blood vessel, for example, the tracer will stop flowing through the body and build up at the site. This can then be investigated by camera. 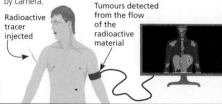 Radioactive tracer injected Tumours detected from the flow of the radioactive material

The process of contaminating substances needs careful consideration as it can be difficult to remove all the contaminating material once it is no longer needed. It is important to consider the half-life of the substances used: it should give enough time to provide the evidence required without causing long-standing contamination.

> Contamination and irradiation can be both useful and problematic.

4 Radioactive contamination

What is meant by irradiation and contamination?

1 Give **one similarity** and **one difference** between irradiation and contamination.

Similarity: ....

....

....

Difference: ....

....

....

Irradiation

2 How is irradiation used to extend the shelf-life of some foods, such as apples?

....

....

....

3 Give **one** medical use for irradiation.

....

Contamination

4 **a)** There is a suspected gas leak in an underground pipe.
Explain how radioactive contamination of the gas supply could help to identify the source of the leak.

....

....

....

....

....

b) Suggest what measures should be taken to ensure that radioactive contamination of the gas supply in part a) does **not** cause unnecessary harm to residents.

....

....

....

....

....

....

Scalar and vector quantities

Physics is concerned with taking measurements of the world around us. Physical quantities are those that can be measured.

There are two types of physical quantity: **scalar** and **vector** quantities.

Scalar quantities have **magnitude** (or size) only.

Energy

Mass

Time

Examples of scalar quantities

Temperature

Speed

Vector quantities have **magnitude** (or size) and **direction**.

Force

Displacement

Velocity

Examples of vector quantities

Acceleration

Momentum

Vector quantities are often represented using arrows:

- The length of the arrow represents the magnitude of the quantity.
- The direction of the arrow represents the direction of the quantity.

Scalar quantities have magnitude only. Vector quantities have magnitude and direction.

For example, a force of 10 N to the left could be represented as an arrow like this:

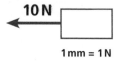

10 N

1 mm = 1 N

Notice the use of a scale in the diagram. This allows large magnitudes to be represented.

Contact and non-contact forces

Force is an example of a **vector** quantity, as forces can act in different directions.

Contact forces occur when two objects are physically in contact with one another.

Examples of contact forces include:
- friction, e.g. between the tyres of a bicycle and the road surface, which will cause the tyres to become warm
- air resistance, e.g. a parachutist falling through the air, which will slow the parachutist down
- normal contact force, e.g. a pencil case resting on the table
- tension, e.g. a string holding up a child's cot mobile, preventing it from falling.

Contact forces occur when two objects are physically in contact with one another, whereas non-contact forces occur without them being physically in contact.

Non-contact forces occur without two objects being physically in contact with one another.

Examples of non-contact forces include:
- gravitational force, which causes masses to be attracted to one another, e.g. the Earth is attracted to the Sun by a gravitational force
- electrostatic force; this exists between charged objects
- magnetic force; this is experienced by any magnetic object when placed in a magnetic field, e.g. opposite poles of a magnet are attracted to one another; like poles repel one another.

Scalar and vector quantities

1 What is meant by a 'scalar' quantity?

....

2 Give **two** examples of scalar quantities.

....

3 What is the difference between a scalar quantity and a vector quantity?

....

....

4 Give **two** examples of vector quantities.

....

....

5 Draw a diagram to represent a force on a box of 40N to the left and 70N to the right.

What is the resultant force acting on the object?

.... N to the ....

Contact and non-contact forces

6 Give **three** examples of contact forces.

....

....

....

7 Give **three** examples of non-contact forces.

....

....

....

5 Gravity

What is gravity?

The Earth's gravitational field is the region around the Earth where objects experience its force of gravity.

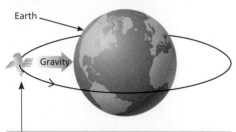

Earth

Gravity

Orbiting satellite experiences an attractive gravitational force towards the Earth. The force of gravity stops it from flying off into outer space.

The Earth's gravitational field strength is defined as the gravitational force on a mass of 1kg. It is given the symbol g and is measured in units of newtons per kilogram (N/kg).

The Earth's gravitational field exerts a force on an apple in a tree. If the apple's stem breaks, the force of gravity causes it to fall faster and faster (until terminal velocity is reached) from the tree to the ground.

When the apple is on the ground, the Earth is still exerting a downward force of gravity on it. However, the apple does not continue to fall because an upward contact force is exerted on it by the ground.

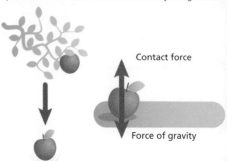

Contact force

Force of gravity

Weight

The force of gravity on an object is also known as its **weight**. The weight of an object can be thought of as acting at a single point: the object's 'centre of mass'.

Weight, mass and gravitational field strength are linked by the following equation:

$$W = mg$$
where
W is the weight, in newtons (N)
m is the mass, in kilograms (kg)
g is the gravitational field strength, in newtons per kilogram (N/kg)

The mass of the satellite in the diagram above is 1200kg. Work out its weight on Earth, given that the gravitational field strength on Earth is 10N/kg.

$W = mg$
$\quad = 1200 \times 10$
$\quad = 12\,000\,N$

At the Earth's surface, the value of g is about 10N/kg. This means that a mass of 1kg experiences a force of gravity of 10N.

Sometimes the mass or the gravitational field strength will not be known, so you will need to rearrange the equation to make the unknown quantity the subject.

If the mass is unknown	$W = mg$ $\dfrac{W}{g} = m$	Divide both sides of the equation by g.
If the gravitational field strength is unknown	$W = mg$ $\dfrac{W}{m} = g$	Divide both sides of the equation by m.

Weight is the force of gravity on an object and can be calculated using the equation $W = mg$

The weight and mass of an object are directly proportional and the weight can be measured using a newtonmeter (also called a spring-balance).

⑤ Gravity

What is gravity?

1 What is meant by the 'gravitational field strength'? State the units in which it is measured.

....

....

2 Draw a diagram to show the direction of the force of gravity on a satellite orbiting the Earth.

Weight

3 Define the term 'weight'. State the units in which it is measured.

....

....

4 In words, write the equation that links gravitational field strength, mass and weight.

....

5 A box has a mass of 5 kg.

Calculate the weight of the box, given that the gravitational field strength on Earth is 10 N/kg.

.... N

6 An astronaut has a mass of 65 kg. On the Moon, the weight of the astronaut is 104 N.

Calculate the gravitational field strength on the Moon.

.... N/kg

⑤ Resultant forces

What are resultant forces?

The movement of an object depends on the forces acting on it. If the forces are equal and opposite, they are balanced and so the resultant force is zero.

If the forces are not equal and opposite, they are unbalanced and the resultant force is not zero.

If the resultant force acting on a **stationary** object is:
- zero, the object will remain stationary

200 N push   Friction 200 N

- not zero, the object will start to move in the direction of the resultant force.

400 N push   Friction 200 N

If the resultant force acting on a **moving** object is:
- zero, the object will continue at the same speed in the same direction

200 N push   Friction 200 N

- not zero, the object will speed up or slow down in the direction of the resultant force.

0 N push   Friction 200 N

> Forces acting on an object can be added or subtracted to give a resultant force.

Free body diagrams

Free body diagrams show the direction and magnitude of all forces acting on an object. The object is shown as a box or a dot.

The free body diagram shows the forces acting on an aeroplane. The length of the arrows is drawn to scale: the longer the arrow, the greater the force. In this example, the weight and the upthrust are equal and the thrust is greater than the air resistance. This means that the aeroplane will move to the right.

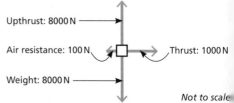
Upthrust: 8000 N

Air resistance: 100 N   Thrust: 1000 N

Weight: 8000 N

Not to scale

Resolving forces

Sometimes it is useful to **resolve** a single force into its two components: the **horizontal** component and the **vertical** component.

A person pulls on a sled with a force of 100 N at an angle of 40° to the horizontal:

In order to resolve the force of 100 N into the horizontal and vertical components, follow these steps:
1. Draw an arrow representing the 100 N force at 40° to the horizontal. Using a scale 1 cm : 10 N, this arrow would be 10 cm long.
2. Draw a horizontal line that ends directly below the 100 N arrow.
3. Draw a vertical line joining the ends of the horizontal line and the 100 N arrow.
4. Add arrowheads to the horizontal and vertical lines.
5. Measure the length of the horizontal and vertical lines.
6. Apply the scale used to draw the original arrow to determine the magnitude of the horizontal and vertical components.

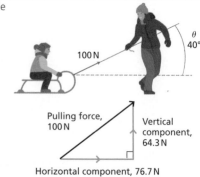
100 N

θ
40°

Pulling force, 100 N

Vertical component, 64.3 N

Horizontal component, 76.7 N

What are resultant forces?

1 Calculate the resultant force acting on the box in the diagram.

5N ← □ → 10N

Resultant force = ... N ....

2 Calculate the resultant force acting on the ball in the diagram.

5N

7N

Resultant force = ... N ....

Free body diagrams

3 Draw a free body diagram to show the forces acting on the boat in the diagram. Label the forces on your diagram.

4 Draw a free body diagram to represent the forces acting on the books in the diagram. Label the forces on your diagram.

Resolving forces

5 A person walking a dog pulls with a force of 50N on the lead, at an angle of 37° to the horizontal.

Using a scale diagram, on a separate piece of paper, determine the horizontal and vertical components of this force.

Horizontal component = ... N

Vertical component = ... N

5 Work done and energy transfer

Work done

When a force moves an object, **work** is done on the object resulting in the **transfer** of **energy**.

In the image, the woman is pushing the wheelbarrow, causing it to move. There is a transfer of energy from the chemical store of the woman to the kinetic store of the wheelbarrow. You can say that **work** has been done on the wheelbarrow.

The work done is equal to the energy transferred. Energy is measured in joules (J), so work done is also measured in joules (J).

work done (J) = energy transferred (J)

Calculating work done

You can calculate the work done on an object using the following equation:

$$W = Fs$$

where

W is the work done, in joules (J)

F is the force, in newtons (N)

s is the distance, in metres (m)

1 joule of work is equal to a force of 1 newton causing a displacement of 1 metre, so $1\,J = 1\,Nm$.

Sometimes the force or the distance of the object will not be known, so you will need to rearrange the equation $W = Fs$ to make the unknown quantity the subject. Remember that when rearranging equations, whatever is done to one side must be done to the other.

If the force is unknown	$W = Fs$ $\dfrac{W}{s} = F$	Divide both sides of the equation by distance.
If the distance is unknown	$W = Fs$ $\dfrac{W}{F} = s$	Divide both sides of the equation by force.

Work done against frictional forces is mainly transformed into heat energy. When work is done on an elastic object to change its shape, the energy is stored in the object as elastic potential energy.

Work done = force × distance, so $1\,J = 1\,Nm$

a) A football is kicked with a force of 10 N. It travels a distance of 15 m.

Calculate the work done on the football as it is kicked.

Begin by listing the information known about the football.

Force = 10 N
Distance = 15 m

Substitute these values into the equation for work done.

$W = Fs$
$W = 10 \times 15$
$W = 150\,J$

b) A man pushes a car with a steady force of 250 N. The car moves a distance of 20 m. How much work does the man do?

Substitute these values into the equation for work done.

$W = Fs$
$W = 250 \times 20$
$W = 5000\,J$ (or $5\,kJ$)

5 Work done and energy transfer

Work done

1 What is meant by the term 'work done'?

....

2 In words, write down the relationship between work done and energy transferred.

....

Calculating work done

3 In words, write the equation that links distance, force and work done. Give the unit for each quantity.

....

....

....

4 A force of 15 N moves a box a distance of 2 m.

Calculate the work done on the box.

.....................................J

5 A wheelbarrow is moved 12 m. 240 J of energy is transferred.

Calculate the force on the wheelbarrow.

.....................................N

6 A car is pushed with a force of 1 kN. 4.5 kJ of energy is transferred.

How far does the car move in metres?

.....................................m

Forces causing a change of shape

As well as causing changes in motion, **forces** can cause a change of shape. Squashing, stretching or bending, or a combination of these, can cause an object to change shape.

When a force deforms an object, the change may be permanent (e.g. squashing a drinks can). This is known as **inelastic deformation**. However, the deformation of other objects may only be temporary (e.g. stretching a band or bending a long bow). They return to their original shape when the force is removed. Objects that return to their original shape after being deformed are made of elastic material. This is known as **elastic deformation**.

Rubber is an elastic material so it can be used to make a catapult:
- The original length of the rubber band on the catapult is 18 cm.
- When stretched, the rubber band is 28 cm long.
- The band's increase in length = 28 – 18 = 10 cm.

This increase in length is also known as the **extension**.

> extension = stretched length – original length

A neck pillow made of polyurethane foam can be used to provide comfort during a long journey:
- Polyurethane foam is an elastic material.
- Your head, resting on the pillow, may cause a compression of about 1 cm, but the pillow returns to its original shape after use.

> compression = original thickness – compressed thickness

Hooke's law

The extension of some elastic materials doubles when the stretching force applied is doubled. These materials are said to obey Hooke's law.

The force and extension are linked by the equation below:

> $F = ke$ where
> F is the force, in newtons (N)
> k is the spring constant, in newtons per metre (N/m)
> e is the extension, in metres (m)

The spring constant tells you how much force is required to extend a material by 1 m.

> Up to the elastic limit, some objects obey Hooke's law, $F = ke$. If the extension-against-force graph for a material is a straight line that passes through the origin (0, 0), the material obeys Hooke's law.

Calculating work done

A force that stretches (or compresses) an elastic object does work, and elastic potential energy is stored in the object. Provided the object is not inelastically deformed, the work done is equal to the elastic potential energy stored in the object.

You can calculate the elastic potential energy stored in an object using the equation below:

> $E_e = \frac{1}{2}ke^2$ where
> E_e is the elastic potential energy, in joules (J)
> k is the spring constant, in newtons per metre (N/m)
> e is the extension, in metres (m)

A spring with a spring constant of 2.5 N/m extends from 20 cm to 40 cm. Calculate the elastic potential energy stored in the spring when it extends.

List the information known about the spring.
Spring constant = 2.5 N/m
Extension = 40 – 20 = 20 cm
Extension = 0.2 m

Substitute these values into the equation for elastic potential energy.
$E_e = \frac{1}{2}ke^2 = \frac{1}{2} \times 2.5 \times 0.2^2$
$E_e = 0.05$ J

5 Forces and elasticity

Forces causing a change of shape

1 The original length of a spring is 20 cm. When a force is applied to the spring, its length changes to 60 cm.

Calculate the extension of the spring in metres.

.... m

2 A sponge is 5 cm thick. When it is squeezed, it becomes 1 cm thick.

Calculate the compression of the sponge.

.... cm

Hooke's law

3 In words, write the equation that links extension, force and spring constant. Give the unit for each quantity.

....

....

4 The graph shows the force and extension for an elastic object.

How does the graph show that the material obeys Hooke's law?

....

....

....

....

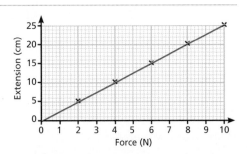

....

Calculating work done

5 An elastic band 10 cm in length is stretched. Its final length is 40 cm. The spring constant of the elastic band is 32 N/m.

Calculate the elastic potential energy stored in the stretched band.

.... J

6 20 J of elastic potential energy are stored in a spring, which has a spring constant of 2.5 N/m.

Calculate the extension of the spring in metres.

.... m

7 A spring has an initial length of 5 cm. 15 J of elastic potential energy are stored in the spring when it is stretched. The spring constant of the spring is 50 N/m.

Calculate the final length of the spring in centimetres. Give your answer to 2 significant figures.

.... cm

Distance and displacement

The **distance** an object travels is the total number of metres (or kilometres or miles) travelled. Distance is a **scalar** quantity as it only has magnitude. In this image, this is shown by the red dotted line.

The **displacement** of an object is the distance moved in a straight line from the start to the end of the journey. Displacement is a **vector** quantity as it has both magnitude and direction. In this image, this is shown by the solid green line.

The difference between distance and displacement

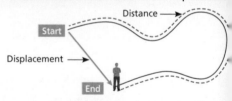

Speed

The **speed** of an object is a measure of how fast it is moving.

To work out the speed of any moving object, you need to know:
- the **distance** it travels
- the **time** it takes to travel that distance.

Some typical values of speed are shown in this table.

Walking	1.5 m/s
Running	3 m/s
Cycling	6 m/s
Sound in air	330 m/s

Calculating distance

Distance can be calculated using the following equation:

$$s = vt \quad \text{where}$$

s is the distance, in metres (m)

v is the speed, in metres per second (m/s)

t is the time, in seconds (s)

A person walks at a speed of 1.5 m/s for 2 hours. Calculate the distance travelled in kilometres.

Begin by listing the information given in the question.

Speed = 1.5 m/s
Time = 2 × 60 × 60 = 7200 s

Substitute these values into the equation for distance.

$s = vt = 1.5 × 7200$
$s = 10800 \text{ m} = 10.8 \text{ km}$

The slope of a **distance–time graph** represents the speed of an object. The steeper the slope, the greater the speed.

The graph below shows:
1 a stationary person
2 a person moving at a constant speed of 2 m/s
3 a person moving at a constant speed of 3 m/s.

N.B. The vertical axis shows distance from a fixed point (0), not total distance travelled.

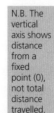
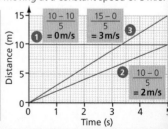

Velocity

Velocity and speed aren't the same thing. The velocity of a moving object describes its speed in a given direction.

Velocity is a vector quantity as it has both magnitude and direction.

Velocity of the car is 40 km/h East

Velocity of the car is 40 km/h South

If an object is travelling at a constant speed in a circle, its direction is constantly changing. This means that the object's velocity is also constantly changing.

If the vehicles on a roundabout travel at a constant speed, the velocity is **not** constant because the direction of the vehicles is constantly changing.

Distance, speed and velocity

Distance and displacement

1 What is the difference between distance and displacement?

....

....

....

Speed

2 State the **two** quantities that are needed in order to determine the speed of an object.

....

3 Complete the table to show typical values of speed for the activities.

Walking	.... m/s
Running	.... m/s
Cycling	.... m/s
Sound in air	.... m/s

Calculating distance

4 In words, write the equation which links distance, speed and time. Give the unit for each quantity in the equation.

....

....

....

5 An athlete runs at a speed of 3.5 m/s for 30 minutes.

How far does the athlete travel in this time?
Give your answer in kilometres.

.... km

6 A cyclist travels 21.6 kilometres in one hour.

Calculate the speed of the cyclist in metres per second.

.... m/s

Velocity

7 Explain why the velocity of an object travelling in a circle is changing, even if the object is travelling at a constant speed.

....

....

....

5 Acceleration

Calculating acceleration

The **acceleration** of an object is the rate at which its velocity changes. It is a measure of how quickly an object speeds up or slows down.

To work out the acceleration of any moving object, you need to know:

- the change in velocity
- the time taken for this change in velocity.

Acceleration can be calculated using the following equation:

$$a = \frac{\Delta v}{t} \quad \text{where}$$

a is the acceleration, in metres per second squared (m/s²)
Δv is the change in velocity, in metres per second (m/s)
t is the time, in seconds (s)

Deceleration is a negative acceleration. It describes an object that is slowing down. It is calculated using the same formula as above.

If the **acceleration is uniform**, the following equation can also be used to calculate acceleration:

$$v^2 - u^2 = 2as$$
$$\text{Rearranging gives: } a = \frac{(v^2 - u^2)}{2s}$$
$$\text{where}$$

v is the final velocity, in metres per second (m/s)
u is the initial velocity, in metres per second (m/s)
a is the acceleration, in metres per second squared (m/s²)
s is the distance, in metres (m)

a) A car starts from rest and accelerates to a velocity of 10 m/s. This takes 5 seconds. Calculate the acceleration of the car.

List the information known about the car.

Initial velocity = 0 m/s
Final velocity = 10 m/s
Velocity change = final velocity – initial velocity
= 10 m/s
Time = 5 seconds

Substitute the values into the equation to calculate acceleration.

$$a = \frac{\Delta v}{t} = \frac{10}{5}$$
$$a = 2 \text{ m/s}^2$$

b) A car starts from rest and accelerates to a velocity of 8 m/s. In this time, the car travels a distance of 15 m. Calculate its acceleration.

List the information known about the car.

Initial velocity = 0 m/s
Final velocity = 8 m/s
Distance travelled = 15 m

Substitute these values into the equation for acceleration.

$$a = \frac{(v^2 - u^2)}{2s} = \frac{(8^2 - 0^2)}{2 \times 15}$$
$$a = 2.13 \text{ m/s}^2$$

Velocity–time graphs

The slope of a **velocity–time graph** represents the acceleration of the object. The steeper the slope, the greater the acceleration.

The area underneath the line in a velocity–time graph represents the total distance travelled.

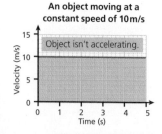
An object moving at a constant speed of 10m/s

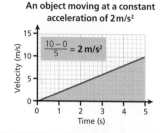
An object moving at a constant acceleration of 2m/s²

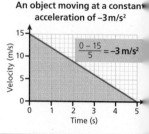
An object moving at a constant acceleration of –3m/s²

The gradient of a velocity–time graph is equal to the acceleration of the object.

Acceleration

Calculating acceleration

1 Define the term 'acceleration' and give the units that it is measured in.

....

....

....

2 In words, write the equation that links acceleration, change in velocity and time.

....

3 A sprinter increases their velocity from 2 m/s to 8 m/s in a time of 6 seconds.

Calculate the acceleration of the sprinter.

.... m/s^2

4 A car accelerates at a rate of 2 m/s^2 for 15 seconds.

If the car starts from rest, what is its final velocity?

.... m/s

Velocity–time graphs

5 State the quantity represented by:

a) the gradient ....

b) the area under a velocity–time graph ....

6 **a)** Plot a velocity–time graph of the data below.

Time (s)	Velocity (m/s)
0	0
5	2
10	4
15	6
20	6

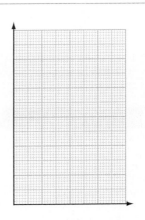

b) Use your graph to determine the acceleration of the object over the first 15 seconds.   .... m/s^2

c) Use your graph to determine the total distance travelled by the object.

.... m

Newton's first law

Newton's first law of motion states that an object will remain stationary or remain travelling at the same speed, in the same direction, unless acted upon by an unbalanced force. The tendency of an object to remain at rest or continue at a constant velocity in the same direction is called **inertia**.

The forces acting on a stationary car are balanced.	If the forces acting on a moving car are equal, the car will remain travelling at the same speed in the same direction.
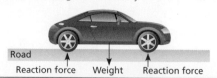 Road Reaction force Weight Reaction force	 5000 N ⟵ ⟶ 5000 N

Newton's second law

Newton's second law states that an object will accelerate when acted on by an **unbalanced** force.

The magnitude of the acceleration is directly proportional to the resultant force acting on the object: $a \propto F$. In other words, if the resultant force doubles, the acceleration will also double.

We can also say that acceleration is inversely proportional to the mass of the object: $a \propto \dfrac{1}{m}$. In other words, if the mass of the object doubles, the acceleration halves.

Putting these two relationships together gives the equation for Newton's second law:

$$F = ma \quad \text{where}$$

F is the force, in newtons (N)
m is the mass, in kilograms (kg)
a is the acceleration, in metres per second squared (m/s²)

Resultant force
→
40 N

Thrust force
100 N

Friction
60 N

The forces on the truck are unbalanced, so it will accelerate in the direction of the resultant force.

A car of mass 1200 kg accelerates at 3 m/s². Calculate the resultant force acting on the car

List the information known about the car.
Mass = 1200 kg   Acceleration = 3 m/s²
$F = ma = 1200 \times 3$ Substitute the values into the
$F = 3600 \text{N}$ equation for force.

Inertial mass is a measure of how hard it is to change an object's velocity. It is given by the ratio of force over acceleration.

Newton's third law

Newton's third law states that for every action force, there is an equal and opposite reaction force. These pairs of forces must:

- act on different bodies
- act in opposite directions
- act along the same line
- be of the same type (e.g. both gravitational forces)
- be of the same magnitude.

The Earth pulls the Moon towards it with a gravitational force. The Moon also pulls the Earth towards it, also with a gravitational force. Notice that the arrows are drawn along the same line and are of equal length. They are drawn pointing in opposite directions.

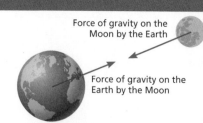
Force of gravity on the Moon by the Earth

Force of gravity on the Earth by the Moon

This shows that the forces act along the same line are equal in magnitude and that the gravitational forces of the Earth and Moon are acting in opposite directions.

5 Newton's laws of motion

Newton's first law

1 State Newton's first law of motion.

_____

_____

_____

2 Use Newton's first law to explain why a book resting on a table is stationary.

_____

_____

_____

_____

Newton's second law

3 State Newton's second law.

_____

_____

_____

4 A lorry with a mass of 20 tonnes accelerates at a rate of $2\,\text{m/s}^2$.

Calculate the force required for this acceleration. 1 tonne = 1000 kg.

_____ N

5 A cyclist and a bike have a combined mass of 100 kg.

If a force of 500 N is applied, calculate the acceleration of the cyclist and bike.

_____ m/s^2

Newton's third law

6 State Newton's third law.

_____

_____

7 List the **five** features of the pairs of forces in Newton's third law.

_____

_____

_____

_____

_____

5 Stopping distance and reaction time

Stopping distance

In order to bring a moving car to stop, the driver must apply the brakes. Along with frictional forces, this causes the car to decelerate.

To apply the brakes, the driver must first **react** to a stimulus, such as a traffic light turning red. The average person has a reaction time of 0.2–0.9 seconds. In this time, the car is still travelling. The distance that the car moves during this time is called the **thinking distance**.

As soon as the driver applies the brakes, the car will begin to decelerate and come to a stop. The distance travelled in the time this takes is called the **braking distance**.

The faster the vehicle is moving, for a given braking force, the longer the braking distance. Learner drivers are required to know the stopping distances of a vehicle at different speeds as given in the Highway Code.

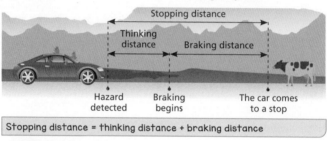

Stopping distance = thinking distance + braking distance

	Thinking distance	Braking distance
	Average car length = 4 metres (13 feet)	
20 mph (32 km/h)	6 m	6 m = 12 metres (40 feet) or 3 car lengths
30 mph (48 km/h)	9 m	14 m = 23 metres (75 feet) or 6 car lengths
40 mph (64 km/h)	12 m	24 m = 36 metres (118 feet) or 9 car lengths
50 mph (80 km/h)	15 m	38 m = 53 metres (175 feet) or 13 car lengths
60 mph (96 km/h)	18 m	55 m = 73 metres (240 feet) or 18 car lengths
70 mph (112 km/h)	21 m	75 m = 96 metres (315 feet) or 24 car lengths

Reaction time and thinking distance

The longer the reaction time, the longer the thinking distance.

The reaction time for a driver can be affected by:
- tiredness – if the driver is tired, their reaction time will increase
- taking drugs – if a driver is taking medication or illegal substances, it may increase reaction time

- drinking alcohol – this also increases reaction time
- distractions, such as the driver using a handheld mobile phone (which is now illegal in the UK) or children playing in the back of a car can increase reaction time.

Factors affecting braking distance

The faster the speed of the vehicle, the greater the braking distance.

Other factors affecting the braking distance include:
- the condition of the road – wet or icy roads can reduce the friction between the tyres and the road surface, increasing the braking distance
- the condition of the vehicle – worn brakes or worn tyres increase the braking distance.

A vehicle travelling at a greater speed requires a greater braking force to bring it to a stop. This means a greater deceleration, which increases the temperature of the brakes owing to increased frictional forces. This can cause the brakes to overheat.

Factors affecting the thinking distance relate to the driver. Factors affecting the braking distance relate to the weather, the vehicle or the road conditions.

⑤ Stopping distance and reaction time

Stopping distance

1 State the average human reaction time. .... s

2 a) What is meant by the term 'thinking distance'?

....

....

b) What is meant by the term 'braking distance'?

....

....

3 The thinking distance for a vehicle travelling at 30 mph is 9 m. The braking distance is 14 m.

Calculate the stopping distance for the vehicle.

.... m

Reaction time and thinking distance

4 Give **three** factors that increase a driver's reaction time.

....

....

....

....

Factors affecting braking distance

5 Give **three** factors that increase the braking distance of a vehicle.

....

....

....

....

6 Explain why a large deceleration can cause the brakes of a vehicle to overheat.

....

....

....

5 Momentum

Momentum

Momentum is a measure of the state of motion of an object. It depends on the **mass** of the object (in kg) and the **velocity** of the object (in m/s).

If a car moves with a greater velocity, it will have more momentum provided its mass hasn't changed. However, a moving truck with a greater mass may have more momentum than the car even if its velocity is less.

Momentum is calculated using the equation below:

$$p = mv$$

where

p is the momentum, in kilogram metres per second (kg m/s)

m is the mass, in kilograms (kg)

v is the velocity, in metres per second (m/s)

a) Calculate the momentum of a car of mass 1200 kg that is moving at a velocity of 30 m/s.

List the information known about the car.

Mass = 1200 kg
Velocity = 30 m/s

Substitute these values into the equation for momentum.

$p = mv = 1200 \times 30 = 36000$ kg m/s

b) A truck has a mass of 4000 kg. Calculate the truck's velocity if it has a momentum of 36 000 kg m/s.

Rearrange the equation to make velocity the subject and substitute the values into it.

$v = \frac{p}{m} = \frac{36000}{4000}$

$v = 9$ m/s

Notice that since the truck has a greater mass than the car in part a), it can move at a slower speed and still have the same momentum.

Conservation of momentum

Provided that two objects are in a **closed system** when they collide, the momentum before the collision will be equal to the momentum after the collision. This is called the law of **conservation of momentum**.

If a moving toy car hits a stationary toy car, both cars will move off together. The direction of the cars will be the same as the direction that the moving car was travelling in. This is because momentum is a vector quantity.

The recoil of a cannon is also an example of the conservation of momentum. As the cannon ball leaves the cannon in the forward direction, the cannon moves backwards to ensure that the momentum before the event is equal to the momentum after the event.

⑤ Momentum

Momentum

1 In words, write the equation that links mass, momentum and velocity.

_____

2 A rugby player has a mass of 80 kg and is running at a velocity of 3 m/s.

Calculate the momentum of the rugby player.

_____ kg m/s

3 A car has a mass of 1500 kg and is moving with a momentum of 15 000 kg m/s.

Calculate the velocity of the car.

_____ m/s

4 A lorry is moving at a speed of 9 m/s with a momentum of 67 500 kg m/s.

Calculate the mass of the lorry.

_____

Conservation of momentum

5 Which has the greater momentum: a 20 g bullet travelling at a speed of 500 m/s or a 50 kg cheetah running at a speed of 30 m/s?

Use calculations to support your answer.

_____

6 State the law of conservation of momentum.

_____

_____

7 Explain what you would observe if a toy car travelling to the right hits a second, stationary toy car.

Use ideas about the conservation of momentum in your answer.

_____

_____

_____

_____

Transverse waves, longitudinal waves and reflection

A wave is a disturbance that travels from one point to another, transferring energy.

Transverse waves (e.g. ripples on a water surface) transfer energy in a direction that is perpendicular (at right angles) to the oscillations (vibrations).

Longitudinal waves (e.g. sound waves) also transfer energy, but the direction of oscillation is parallel to the direction of energy transfer.

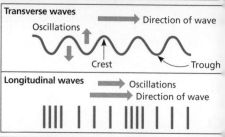

Transverse waves

Oscillations — Direction of wave
Crest — Trough

Longitudinal waves — Oscillations — Direction of wave

Wave	Transverse	Longitudinal	Carries energy	Can travel through a vacuum	Can be reflected
Light	✓	✗	✓	✓	✓
Sound	✗	✓	✓	✗	✓
Water surface	✓	✗	✓	✗	✓

When light strikes a surface, it changes direction. This is called **reflection**.

The **normal** is a line that is perpendicular to the reflecting surface at the point of incidence. The normal is used to calculate the angles of incidence and reflection.

Angle of incidence = Angle of reflection

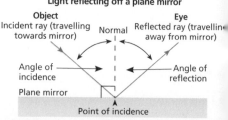

Light reflecting off a plane mirror

Object — Incident ray (travelling towards mirror) — Normal — Eye — Reflected ray (travelling away from mirror)

Angle of incidence — Angle of reflection

Plane mirror — Point of incidence

Wave terms and calculations

Amplitude: the maximum displacement of a point on a wave away from its undisturbed position (i.e. the distance in metres from the undisturbed position to the crest, or from the undisturbed position to the trough)

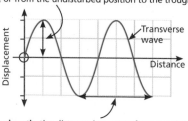

Displacement — Distance — Transverse wave

Wavelength: the distance in metres from a point on one wave to the equivalent point on the next wave

The **frequency** is measured in hertz, where 1 hertz means one complete wave passes a given point every second. The **time period** is the time taken for one wave to pass a given point.

$T = \dfrac{1}{f}$ which can be rearranged to $f = \dfrac{1}{T}$

where

T is the time period, in seconds (s)
f is the frequency, in hertz (Hz)

The equation below calculates the speed of a wave

$v = f\lambda$ where

v is the wave speed, in metres per second (m/s)
f is the frequency, in hertz (Hz)
λ is the wavelength, in metres (m)

a) Calculate the frequency of a wave with a time period of 5 seconds.

$f = \dfrac{1}{T} = \dfrac{1}{5}$
$= 0.2\,\text{Hz}$

b) Calculate the speed of a wave with a wavelength of 50 cm and a frequency of 10 Hz.

$v = f\lambda = 10 \times 0.5$ Convert the wavelength to metres.
$v = 5\,\text{m/s}$

Types and properties of waves

Transverse waves, longitudinal waves and reflection

1 Give an example of:

a) a transverse wave ....

b) a longitudinal wave ....

2 Label the diagram with the words 'trough' and 'crest'.

Direction of
travel of the wave

3 A ray of light is incident on a plane mirror. The angle of incidence is 35°.

What is the angle of reflection of the light ray? .... °

Wave terms and calculations

4 What is meant by the 'amplitude' of a wave?

....

....

5 Define the 'frequency' of a wave and state the unit in which it is measured.

....

....

6 A wave has a time period of 10 seconds.

Calculate the frequency of the wave.

.... Hz

7 In words, write the equation that links frequency, wavelength and wave speed.

....

8 A wave with a frequency of 500 Hz has a wavelength of 0.1 cm.

Calculate the speed of the wave in metres per second.

.... m/s

9 A wave is travelling at 3×10^6 m/s with a frequency of 3000 Hz.

Calculate the wavelength of the wave. Give your answer in standard form.

.... m

Electromagnetic waves

Electromagnetic radiations are disturbances in an electric field. They travel as waves and transfer energy from one place to another.

All types of electromagnetic radiation travel at the same speed through a vacuum.

Each type of electromagnetic radiation has a **different wavelength** and a **different frequency**. Together, they form the **electromagnetic spectrum**, shown below.

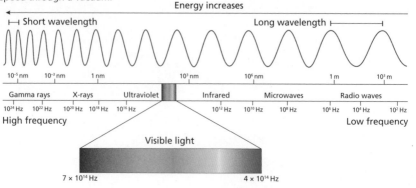

When radiation is absorbed by a substance:
- the energy is absorbed and makes the substance heat up
- it may create an alternating current of the same frequency as the radiation.

> Different wavelengths of electromagnetic radiation are **reflected**, **absorbed** or **transmitted** in different ways by different substances and surfaces.

Visible light is the only part of the electromagnetic spectrum that can be seen by the human eye.

White light is made up of different colours. The colours are **refracted** by different amounts as they pass through a prism:
- Red light is refracted the least.
- Violet light is refracted the most.

Properties of electromagnetic waves

Electromagnetic waves refract when they cross the boundary between two transparent materials. They change direction owing to a change in speed.

The diagrams show an experiment in which a red laser beam is directed at a glass block. Light is a transverse wave, so the laser beam can be represented by rows of crests.

Air | Glass block

Laser beam changes direction when it travels from the air into the glass.

Laser beam bends back to the original direction when it comes out of the glass into the air.

As the laser beam enters the glass, its speed decreases. The part of one row of crests in the glass moves more slowly than another part that is still in the air, so the beam bends.

Air Glass

Light travels more slowly in glass than air because glass is much denser. When the laser beam re-enters the air, it speeds up again.

Radiation type	Property
Radio waves	If radio waves are absorbed, they may create an alternating current with the same frequency of the radio waves. This means they can induce oscillations in an electrical circuit.
Ultraviolet (UV) waves	UV waves can cause premature ageing of the skin and skin cancer.
UV, X-rays and gamma rays	These waves are all hazardous to the human body.
X-rays and gamma rays	These waves are ionising, which means they can cause mutations in body cells.

Types and properties of electromagnetic waves

Electromagnetic waves

1 List the types of electromagnetic wave from longest wavelength to shortest wavelength.

....

....

2 Calculate the speed of blue light, given the following information.

Frequency = 7.0×10^{14} Hz

Wavelength = 4.3×10^{-7} m

Speed = frequency × wavelength

.... m/s

Properties of electromagnetic waves

3 Define the term 'refraction'.

....

....

4 List **three** parts of the electromagnetic spectrum that are hazardous to the human body.

....

5 Use the diagram to explain how radio waves can be used to induce electrical oscillations in an electrical circuit.

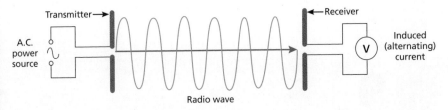

....

....

6 X-rays and gamma rays can both cause cancer in the human body. Explain how.

....

....

....

....

Uses and applications of electromagnetic waves

Uses and applications of electromagnetic waves

Electromagnetic waves	Uses	Effects
Radio waves	• Transmitting radio and TV signals between places across the Earth.	• High levels of exposure for short periods can increase body temperature, leading to tissue damage.
Microwaves	• Satellite communication networks and mobile phone networks (they can pass through the Earth's atmosphere). • Cooking – water molecules absorb microwaves, and heat up.	• May damage or kill cells because microwaves are absorbed by water in the cells, which heat up.
Infrared rays	• Grills, toasters and radiant heaters (e.g. electric fires). • Remote controls for televisions, etc. • Optical fibre communication.	• Absorbed by skin and felt as heat. • An excessive amount can cause burns.
Ultraviolet rays	• Security coding – special paint absorbs UV and emits visible light. • Sun tanning and sunbeds.	• Passes through skin to the tissues below. • High doses can kill cells. • A low dose can cause cancer.
X-rays	• Producing shadow pictures of bones and metals. • Treating certain cancers.	• Passes through soft tissues (some is absorbed). • High doses can kill cells. • A low dose can cause cancer.
Gamma rays	• Killing cancerous cells. • Killing bacteria on food and surgical instruments.	• Passes through soft tissues (some is absorbed). • High doses can kill cells. • A low dose can cause cancer.

Different parts of the electromagnetic spectrum have different uses. Radio waves and microwaves can be used for communication. X-rays and gamma rays can be used for medical imaging and treatment.

Uses and applications of electromagnetic waves

Uses and applications of electromagnetic waves

1 Draw lines to match the type of electromagnetic wave to its use.

Type of electromagnetic wave	Use
Radio waves	Remote controls for televisions
Microwaves	Transmitting TV signals
Infrared radiation	Producing shadow pictures of bones
Ultraviolet radiation	Satellite communications
X-rays	Killing bacteria on food
Gamma rays	Security coding

2 Explain why exposure to microwaves can damage cells in the body.

....

....

3 How are X-rays and gamma rays hazardous to the human body?

....

....

....

....

....

4 Sunbeds in tanning studios use ultraviolet radiation.

Explain why care must be taken when using sunbeds.

....

....

....

....

Poles and magnetic fields

A **magnet** is something that creates an invisible **magnetic field** in the space that surrounds it. All magnets have a north pole and a south pole.

Opposite poles of two magnets attract	Like poles of two magnets repel
The attractive force of two opposite poles is caused by the interaction of their fields and is called a **magnetic force**. Since it exists without the magnets touching, it is a non-contact force. 	The magnetic force experienced between two north poles, or between two south poles, is a repulsive, non-contact force. 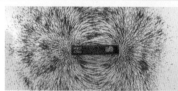

Iron filings can be used to show a magnetic field because iron is a magnetic material. They are pushed and pulled into the shape of the magnetic field. Other magnetic materials include nickel, steel, cobalt and neodymium.

When a magnetised steel needle is placed on a piece of cork floating in a bowl of water, the needle's magnetic field interacts with the Earth's magnetic field. This makes the needle rotate so that one end points towards the Earth's North Pole. This end is called a north-seeking pole.

A suspended magnetised needle forms part of a compass. A navigation compass is used by sailors and hikers to find their way.

A smaller version of a compass, called a **plotting compass**, can be used to find out more about the shape of the magnetic field of a bar magnet. The red half of the needle gives the direction that the compass is pointing.

In a diagram, the magnetic field is represented by lines with arrows showing the direction a compass would point when placed at that position. Notice how the arrows on the magnetic field lines always point from north to south.

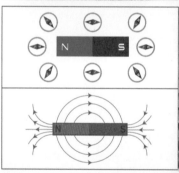

Types of magnetism

Permanent magnets keep their magnetic properties. Most contain the strongly magnetic materials of iron, nickel or cobalt.

Repeatedly stroking a steel nail with the north pole of a magnet in the direction shown by the arrows magnetises the nail. In this case, the point of the nail becomes a north pole, and the head becomes a south pole. A magnetised steel nail will retain its magnetism unless it is dropped or hit with a hammer.

Nail made of steel
(iron mixed with carbon)

Inside the nail, each atom of iron acts like a tiny magnet. The atoms band together in a region within the nail, so their tiny magnets create magnetic fields with the same direction. These regions are called **domains**.

Stroking a steel nail with a magnet aligns its domains, giving it permanent magnetism.

Small section of nail before (left) and after (right) being magnetised

Induced magnets are made from materials that become magnetic when placed in a magnetic field. There is always a force of attraction between a magnet and an induced magnet. Induced magnets lose their magnetism when removed from the magnetic field.

Permanent magnets produce their own magnetic fields. Induced magnets only become magnetic when placed in a magnetic field.

7 Magnetism

Poles and magnetic fields

1 What are the **two** poles of a magnet called?

....

2 Magnetism is an example of a non-contact force.

Explain how this can be shown using two bar magnets.

....

....

....

....

....

....

3 What is meant by the term 'magnetic field'?

....

....

4 Iron is a magnetic material.

Name **two** other magnetic materials.

....

....

Types of magnetism

5 What is a permanent magnet? Give an example.

....

....

6 Explain how a steel nail can become magnetised.

....

....

....

7 What is an induced magnet?

....

....

Temporary magnets

A weak magnetic field can be created with a piece of wire and a battery. The wire is made of copper with a plastic coating.

When the switch in the circuit is open, no electric current flows and a magnetic field is not created so the plotting compass needles point north. 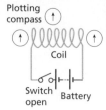	When the switch is closed, an electric current is flowing in the coil. The plotting compass needles have changed direction: the coil is now producing a magnetic field. It is an **electromagnet**. 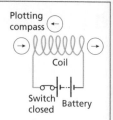

A very strong magnetic field, which can be switched on and off as desired, can be made with a long coil containing a core of soft iron. It is called a **solenoid**.

An iron-cored solenoid, carrying an electric current, creates a strong magnetic field.

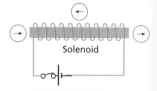

Magnet	Type of magnet	Strength of field
Bar magnet	Permanent	Quite strong
Electromagnet with air core	Temporary	Weak
Solenoid with iron core	Temporary	Very strong

Advantages and applications of electromagnets

Most useful electromagnets consist of a coil, sometimes containing a soft iron core. When a battery is connected to the coil, it drives an electric current around it. This creates the magnetic field.

Three advantages of an electromagnet compared with a permanent magnet are:
- the magnetic field can be easily switched on or off by connecting or disconnecting the battery
- the electric current can be varied so the strength of the field can be controlled
- the direction of the field can be easily reversed by reversing the battery connections.

Some of the uses of electromagnets include:
- headphones
- loudspeakers
- electric bells
- scrapyard cranes
- data storage
- magnetic locks.

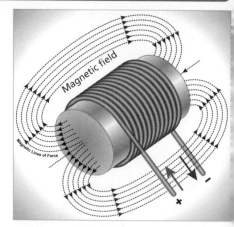

The magnetic field of an electromagnet can be switched on or off, can be made stronger or weaker, and made to change direction

7 Electromagnetism

Temporary magnets

1 What is an electromagnet?

_____

_____

_____

2 What is a solenoid?

_____

3 Complete the table.

Magnet	Type of magnet	Strength of field
Bar magnet		Quite strong
Electromagnet with air core	Temporary	
Solenoid with iron core		Very strong

Advantages and applications of electromagnets

4 List **three** advantages of electromagnets compared with a permanent magnet.

_____

_____

_____

_____

_____

5 List **three** applications of electromagnets.

_____

_____

_____

The motor effect and factors affecting the force on a wire

The motor effect uses current to produce movement.

When a conductor (wire) carrying an electric current is placed in a magnetic field, the magnetic field formed around the wire interacts with the permanent magnetic field. This causes the wire to experience a force, which makes it move.

> The wire will **not** experience a force if it is parallel to the magnetic field.

How to increase the size of the force on the wire	How to reverse the direction of the force on the wire
• **Increase the size of the current** (e.g. have more cells) 	• **Reverse the direction of flow** of the current (e.g. turn the cell around) 
• **Increase the strength of the magnetic field** (e.g. have stronger magnets) 	• **Reverse the direction of the magnetic field** (e.g. swap the position of the north and south poles) 

Fleming's left-hand rule

Fleming's left-hand rule can be used to determine the relative orientation of the force, the magnetic field and the current on the wire in a magnetic field.

Hold your thumb, first finger and second finger at right angles to one another:

- Your thumb will point in the direction of the force (and so direction of movement) on the wire.
- Your first finger needs to be lined up with the magnetic field, from north to south.
- Your second finger needs to be lined up with the current, from positive to negative.

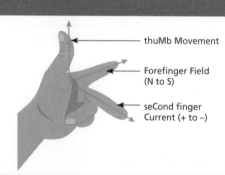

thuMb Movement

Forefinger Field (N to S)

seCond finger Current (+ to −)

Calculating the force on a current-carrying wire

For a current-carrying wire at right angles to a magnetic field:

> $F = BIL$ where
> F is the force, in newtons (N)
> B is the magnetic flux density, in teslas (T)
> I is the current, in amperes (A)
> L is the length of the wire, in metres (m)

A current of 3A flows through a 1.5m long wire when it is placed in a magnetic field of flux density 0.5T. Calculate the force on the wire.

Substitute the given values into the formula for force.

$F = BIL = 0.5 \times 3 \times 1.5$

$F = 2.25\,\text{N}$

You may need to rearrange the equation $F = BIL$ to make the unknown quantity the subject:

If the magnetic flux density is unknown		If the current is unknown		If the length of wire is unknown	
$F = BIL$		$F = BIL$		$F = BIL$	
$\dfrac{F}{IL} = B$	Divide both sides of the equation by IL.	$\dfrac{F}{BL} = I$	Divide both sides of the equation by BL.	$\dfrac{F}{BI} = L$	Divide both sides of the equation by BI.

Fleming's left-hand rule

The motor effect and factors affecting the force on a wire

1 What is meant by the 'motor effect'?

....

....

....

....

....

2 State **two** ways in which the force on a wire in a magnetic field can be increased.

....

....

Fleming's left-hand rule

3 State **two** ways in which the direction of force on a wire on a magnetic field can be reversed.

....

....

4 The diagram shows a current-carrying wire placed in a magnetic field.

Use Fleming's left-hand rule to determine the direction that
the wire will move.

....

....

....

....

....

Calculating the force on a current-carrying wire

5 In words, write the equation that links current, force, length and magnetic flux density.

....

6 Calculate the force on a 2.5 m current-carrying wire that is placed in a magnetic field of
flux density 1.5 T with a current of 10 A flowing through it.

.... N

7 Electric motors and loudspeakers

Interacting magnetic fields

When two magnets are moved close enough for their magnetic fields to overlap, an interaction occurs. The interaction can result in either an attractive or a repulsive force.

> Motion caused by the interaction between a permanent magnet and an electromagnet is called the **motor effect**.

A similar interaction occurs between a magnet and a magnetic field produced by a wire carrying an electric current.

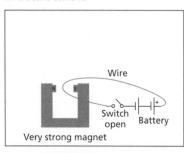

Wire
Switch open   Battery
Very strong magnet

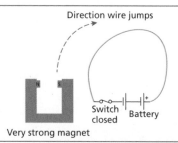

Direction wire jumps

Switch closed   Battery
Very strong magnet

When the switch is closed, electric current flows in the wire. The wire instantly jumps out of the magnet in the direction shown by the dashed red arrow.

The electric motor

The function of an electric motor is to create rotation. A battery (not shown) is connected to the coil, which becomes an electromagnet.

- The coil's magnetic field and the field produced by the magnets interact.
- The interaction creates an upward force on the left side of the coil. This is shown by the red arrow pointing upwards.
- The interaction also creates a downward force on the right side of the coil. This is shown by the red arrow pointing downwards.
- The upward and downward forces, shown by the red arrows, make the coil rotate clockwise.

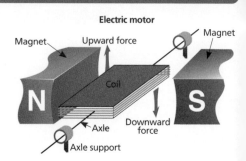

Electric motor

Magnet   Upward force   Magnet
N   Coil   S
Axle   Downward force
Axle support

Household appliances that are fitted with an electric motor include: washing machine; tumble dryer; microwave oven; electric tin opener; electric food mixer; electric fan.

On page 280, we saw that the force on the wire in an electric motor can be increased.

This can be done by increasing the size of the current flowing through the wire or by increasing the strength of the magnetic field.

The strength of the magnetic field is given by the magnetic flux density. The higher the magnetic flux density, the greater the strength of the magnetic field.

These are linked by the equation:

$$F = BIL$$

where

F = force on the wire (N)
B = magnetic flux density (T)
I = current (A)
L = length (m)

Interacting magnetic fields

1 The motor effect is the motion caused by the interaction between a permanent magnet and what?

....

The electric motor

2 The diagram shows an electric motor. Explain how an electric motor works.

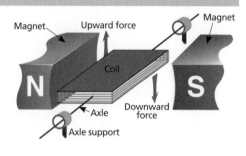

Magnet · Upward force · Magnet · Coil · N · S · Axle · Downward force · Axle support

....

....

....

....

....

....

3 Name **three** household appliances that contain an electric motor.

....

....

4 Assuming that the length of, and current flowing through, a wire in an electric motor remain constant, if the magnetic flux density is doubled, what effect would this have on the force on the wire?

....

....

5 The current flowing through a 2 m length of wire is 5 A. The wire is placed in a magnetic field with a magnetic flux density of 2.5 T.

Giving each answer to 2 significant figures, calculate:

a) the force on the wire

....

b) the force on the wire if the current were halved.

....

Mixed questions (paper 1)

1 A student investigates the digestion of lipids by lipase.

They find that when the lipid is digested, the pH of the mixture falls to pH 5.

a) Explain why the digestion of lipid causes the pH to fall.

....

....

....

b) The student sets up three different test tubes for the experiment.

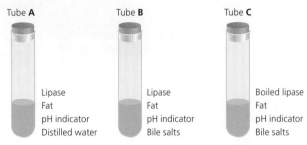

Tube **A**

Lipase
Fat
pH indicator
Distilled water

Tube **B**

Lipase
Fat
pH indicator
Bile salts

Tube **C**

Boiled lipase
Fat
pH indicator
Bile salts

They find that the pH indicator does not change colour in tube **C**.

Explain why this is.

....

....

....

2 Some cells are specialised to carry out a specific function.

Draw **one** line from each description to the correct function of that type of cell.

Description	Function
A cell that is hollow and forms tubes	To contract
A cell that has a flagellum	To transport water
A cell that is full of protein fibres	To carry nerve impulses
A cell that has a long projection with branched endings	To swim

Mixed questions (paper 1)

3 The human body has a double circulatory system.

Complete the sentences.

Oxygenated blood from the .... returns to the left atrium of the

heart in the pulmonary .... From here, it enters the left ventricle

and leaves the heart via the .... to go to the body. From the body,

.... blood returns via the .... to the right

atrium and then leaves the heart in the pulmonary artery to go to the lungs.

4 Plants have two separate transport tissues.

a) Give the name of the tissue responsible for transporting water and mineral ions.

....

b) Give the name of the tissue responsible for transporting dissolved sugars.

....

5 Complete the word equation for aerobic respiration.

glucose + ... → .... + water

6 Complete the table with a tick or cross to show if the structures are present or absent in the cells listed.

Type of cell	Nucleus	Cytoplasm	Cell membrane	Cell wall
Plant cell				
Bacterial cell				
Animal cell				

7 **a)** Give **two** differences between osmosis and diffusion.

....

....

....

....

....

b) Give **two** differences between diffusion and active transport.

....

....

....

....

....

Mixed questions (paper 1)

8 Chlorine is a Group 7 element.

What is the name given to the elements in Group 7? Tick one box.

Halogens ☐

Transition metals ☐

Noble gases ☐

Alkali metals ☐

9 Chlorine has an atomic number of 17.

Draw the electronic structure of a chlorine atom.

10 a) Chlorine can react with hydrogen to form hydrogen chloride.

Name the bonding present in hydrogen chloride.

....

b) Give the electronic structure of a chloride ion.

....

c) Sodium can react with chlorine to form sodium chloride. Describe the bonding in sodium chloride.

....

....

11 Explain why the melting point of hydrogen chloride is lower than the melting point of sodium chloride.

....

....

....

....

....

....

12 a) Ammonia can react with nitric acid to make ammonium nitrate, NH_4NO_3.

Calculate the relative formula mass of ammonium nitrate.

A_r of H = 1, N = 14, O = 16

b) Calculate the percentage composition of nitrogen in ammonium nitrate.

13 a) Define the term 'oxidation'.

b) Write a word equation to show the oxidation of magnesium to form magnesium oxide.

14 Copper sulfate can undergo a chemical reaction with magnesium.

a) Give the name of the substance that is being oxidised.

b) Give the name of the substance that is being reduced.

c) Write a word equation for this reaction.

15 Combustion of natural gas is used in our homes for heating and cooking.

Draw an energy level diagram for the complete combustion of methane to form carbon dioxide and water.

Show on your diagram the activation energy and the energy given out in the reaction.

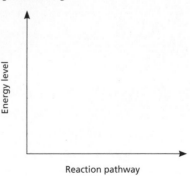

Energy level

Reaction pathway

Mixed questions (paper 1)

16 A golfer strikes a golf ball and it travels at a speed of 90 m/s. It gains 203 J of kinetic energy.

Calculate the mass of the golf ball in grams.

.... g

17 An apple is hanging from a tree at a height of 2.5 m. The mass of the apple is 250 g.

Taking the gravitational field strength to be 9.8 N/kg, calculate the gravitational potential energy of the apple. Give your answer to 2 significant figures.

....

18 An electric hairdryer is connected to the mains supply and has a resistance of 100 Ω. The current flowing through the hairdryer is 5 A.

Calculate the power of the hairdryer.

.... W

19 Hydroelectricity is a renewable energy resource that uses the flow of moving water to generate electricity.

Give **two advantages** and **two disadvantages** of using hydroelectric as an energy resource.

Advantages: ....

....

Disadvantages: ....

....

20 A motor runs for 15 minutes and transfers 4500 C of charge.

Calculate the electrical current flowing through the motor.

....

Mixed questions (paper 1)

21 A student needs to charge their camera. They connect the camera charger to the mains power supply, with a potential difference of 230 V.

If the charging current is 0.1 A, calculate the resistance in the circuit as the camera charges.

.... Ω

22 Draw **one** line from each component to the correct current–potential difference graph.

Resistor	Filament lamp	Diode

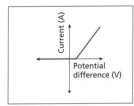 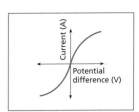 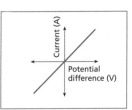

23 A 900 W microwave oven transfers 135 kJ of energy when it is turned on.

For how long is the microwave oven turned on? Give your answer in seconds.

.... s

24 Balance the nuclear equation.

$$_{53}^{}\text{I} \rightarrow {}^{131}\text{Xe} + {}_{-1}^{}\text{e}$$

25 450 kJ of energy are supplied to a 2.0 kg brick. Brick has a specific heat capacity of 840 J/kg°C.

Calculate the change in temperature of the brick. Give your answer to 2 significant figures.

.... °C

Mixed questions (paper 2)

1 The diagram shows several interconnected food chains from a habitat.

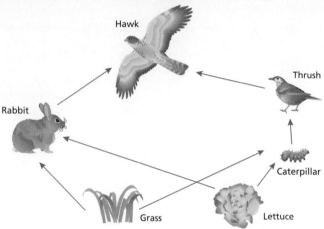

Hawk

Thrush

Rabbit

Caterpillar

Grass   Lettuce

a) Draw one food chain containing four organisms from this habitat.

....

b) Complete this table by writing in the number of each type of organism found in the diagram. One has been done for you.

Type of organism	Number in the diagram
Producers	2
Herbivores	
Prey	
Apex predators	

2 The diagram shows the formation of a zygote and an embryo.

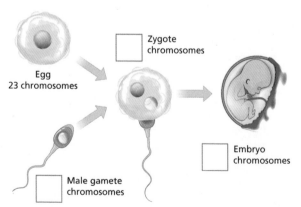

Zygote chromosomes

Egg
23 chromosomes

Embryo chromosomes

Male gamete chromosomes

On the diagram, write the number of chromosomes in the male gamete, the zygote and the embryo cell.

Mixed questions (paper 2)

3 Reflex actions are designed to prevent the body from being harmed. A person receives a sharp blow just below the kneecap and this causes the leg to straighten rapidly.

The person concentrates and tries to stop their leg straightening when it is hit.
However, they cannot stop the movement. Explain why.

....

....

....

....

4 The quagga used to live in South Africa. The last quagga died in 1883. Its scientific name was *Equus quagga*.

a) What genus did the quagga belong to?

....

b) Scientists analysed DNA from the quagga and concluded that it is the same species as a type of zebra that is alive today. The scientific name of the zebra alive today is *Equus burchellii*.

Why did scientists need to change the scientific name of either the quagga or the zebra, but did not need to change their common names?

....

....

....

5 Scientists want to produce partially striped zebras from fully striped zebras using selective breeding.

Describe the steps in this process.

....

....

....

....

....

5 Doctors are trying to find a cure for a human disease by inserting a working gene.
Put a ring around the name for this type of process.

cloning	genetic engineering	natural selection	selective breeding

Mixed questions (paper 2)

7 Define the term 'activation energy'.

....

....

8 A student wanted to investigate how the concentration of hydrochloric acid affected the rate of reaction with magnesium ribbon.

The figure shows the equipment that the student used.

a) Give the independent variable for this experiment.

....

b) Give the unit of the dependent variable of this experiment.

....

c) Predict the effect of changing concentration of acid on the rate of reaction.

Use ideas from collision theory in your answer.

....

....

....

....

....

....

9 Crude oil is a mixture of alkanes.

The figure below is an example of an alkane found in crude oil.

$$H-\overset{\overset{\displaystyle H}{|}}{\underset{\underset{\displaystyle H}{|}}{C}}-\overset{\overset{\displaystyle H}{|}}{\underset{\underset{\displaystyle H}{|}}{C}}-H$$

a) Name the alkane in the figure above.

....

b) Give the general formula for an alkane.

....

10 A student analysed a colourless solution to determine the ions present.

a) The student completed a flame test on the solution. The flame turned orange-red.

Give the formula of the ion present.

b) The student added some sodium hydroxide solution to a sample of the solution. Use your answer to part **a)** and suggest the observation made.

11 The atmosphere is an envelope of gas that surrounds our planet.

a) Give the name of the main element in dry air.

b) Give the formula of the substance that makes up $\frac{1}{5}$ of dry air.

12 For how long has the composition of the atmosphere been stable? Tick one box.

200 million years ☐

4.6 billion years ☐

2.7 million years ☐

200 billion years ☐

13 Name the atmospheric pollutant that can cause global dimming.

14 a) Alkanes are hydrocarbons. Define the term hydrocarbon.

b) What process is used to make alkenes?

15 An astronaut has a mass of 70 kg. On the Moon, their weight is 112 N.

Calculate the gravitational field strength on the Moon.

_____ N/kg

Mixed questions (paper 2)

16 A gardener pushes a wheelbarrow with a force of 50 N. 0.5 kJ of energy is transferred.

How far does the wheelbarrow move?

.... m

17 Red light travels at a speed of 3×10^8 m/s with a frequency of 4×10^{14} Hz.

Calculate the wavelength of red light in metres. Give your answer in standard form.

.... m

18 A student makes a journey of 15.5 km in a time of 30 minutes.

Calculate their average speed, in metres per second. Give your answer to 2 significant figures.

.... m/s

19 An object accelerates at a rate of 2 m/s² to a final velocity of 10 m/s in a time of 3 seconds.

Calculate the initial velocity of the object.

.... m/s

20 A tractor has a mass of 2500 kg and a momentum of 11 250 kg m/s.

Calculate the velocity of the tractor.

.... m/s

21 40 J of elastic potential energy is stored in a stretched spring, with a spring constant of 4.5 N/m.

Calculate the extension of the spring. Give your answer to 2 significant figures.

.... m

Mixed questions (paper 2)

22 Three uses of electromagnetic radiation are listed below. For each one, state the type of electromagnetic wave that is used for the given application.

a) Transmitting television signals ....

b) Satellite communications ....

c) Remote controls for televisions ....

23 In which direction do magnetic field lines always point? ....

24 Electromagnets are made from a coil of wire. An iron core is often added.

Why is the iron core added?

Soft iron core Turns of insulated copper wire N S

....

....

25 The generator effect uses the movement of a magnet relative to a coil of wire to produce a current in the wire. Moving the magnet into the coil induces a current in one direction.

State **one** way in which the direction of the current could be reversed.

....

....

26 A transformer has 100 turns on the primary coil. A potential difference of 230 V is applied to the primary coil and a potential difference of 4500 V is produced on the secondary coil.

a) How many turns are there on the secondary coil?

.... turns

b) Is the transformer a **step-up** or a **step-down** transformer? Give a reason for your answer.

....

....

....

27 A current of 13 A flows through a 0.5 m long wire when it is placed in a magnetic field of flux density 0.75 T.

Calculate the force on the wire. Give your answer to 2 significant figures.

.... N

Required practical 1

Microscope drawings

Aim: To use a microscope to make observations of biological specimens and produce labelled scientific drawings.

Use a light microscope to observe, draw and label a selection of plant and animal cells.

Sample method	Considerations, mistakes and errors
1. Place a tissue sample on a microscope slide. 2. Add a few drops of a suitable stain. 3. Lower a coverslip onto the tissue. 4. Place the slide on the microscope stage and focus on the cells using low power. 5. Change to high power and refocus. 6. Draw any types of cells that can be seen. 7. Add a scale line to the diagram.	• The scale line can be added by focusing on the millimetre divisions of a ruler.

Key points:

- The stain used depends on the tissue but iodine solution may be used for plant cells as it stains starch. Methylene blue might be used to stain the nucleus.
- Look very closely at a specimen to draw it accurately.
- Drawings should always be in pencil. Fine detail cannot be drawn accurately unless the pencil has been sharpened.
- The outlines of any structures should be drawn but there should be no colouring or shading.
- Do not draw sketchy lines.
- If required, the drawing should be labelled, using label lines that do not cross or obscure the drawing.

Expected results:

- The drawing should accurately show individual cells or the outline of tissues – whichever is required.

A light microscope

Required practical 2

Osmosis

Aim: To investigate the effect of a range of concentrations of salt or sugar solutions on the mass of plant tissue.

REQUIRED PRACTICAL	
Investigating the effect of different concentrations of sugar solution on plant tissue.	
Sample method	**Considerations, mistakes and errors**
Potatoes can be used to measure the effect of sugar solutions on plant tissue: 1. Cut some cylinders of potato tissue to the same length and measure their mass. 2. Place the cylinders in different concentrations of sugar solution. 3. After about 30 minutes, remove the cylinders and measure their mass again. If the cylinders change in mass, they have gained or lost water by osmosis.	• The cylinders need to be left in the solution long enough for a significant change in mass to occur. • Before the mass of the cylinders is measured again, they should be rolled on tissue paper to remove any excess solution.
Variables	**Hazards and risks**
• The independent variable is the one deliberately changed – in this case, the concentration of sugar solution. • The dependent variable is the one that is measured – in this case, the change in mass of the potato. • The control variables are kept the same – in this case, the temperature, the length of time the cylinders were left in the solution, and volume of the solution.	• Care must be taken when cutting the cylinders of potato.

Key points:

- The potato cylinders will gain or lose in mass if they take up or lose water by osmosis.
- The results depend on the difference between the concentration of the cytoplasm of the potato cells and the concentration of the sugar solution.
- It is important to calculate the percentage change in mass from the results because the starting masses of the potato cylinders will probably be slightly different.

Expected results:

- In a dilute sugar solution, the potato cylinders will gain water by osmosis and so their mass will increase.
- In a concentrated sugar solution, the potato cylinders will lose water by osmosis and so their mass will decrease.
- The concentration of the cytoplasm of the potato cells is equivalent to the concentration of the solution when there is no change in mass. This can be read off a graph of the results.

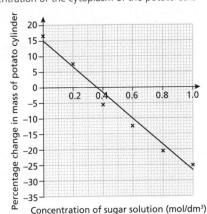

Required practical 3

Testing for biological molecules

Aim: Use qualitative reagents to test for a range of carbohydrates, lipids and proteins.

REQUIRED PRACTICAL	
Use qualitative reagents to test for a range of carbohydrates, lipids and proteins.	
Sample method	**Considerations, mistakes and errors**
1. To test for sugars, e.g. glucose, add Benedict's reagent and heat in a water-bath for two minutes. If sugar is present, it will turn red. 2. To test for starch, add iodine solution. If starch is present, it will turn blue-black. 3. To test for protein, add biuret reagent. If protein is present, it will turn purple. 4. To test for lipids, shake with ethanol (alcohol) in a test tube. Then add a few drops of the alcohol to a test tube of water. If there is lipid present then a milky-white emulsion will form.	• Do not boil the mixture for a long time, because any starch present might break down into sugar and test positive. • Refer to 'iodine solution' not 'iodine'. • Sometimes the purple colour is difficult to see. Try holding the test tube in front of a sheet of white paper.
	Hazards and risks • In the Benedict's test use a water-bath to heat the test tube. • When doing the emulsion test make sure there are no flames nearby.

Key points:

- Starch and glucose are both carbohydrates but different tests are used to detect them.
- The test for sugars with Benedict's reagent is the only test that needs heating.
- The tests give information about whether the substance is present or not. This is called quantitative.
- The colour of Benedict's reagent will change from blue to green, to yellow and finally red depending on how much sugar is present. This can give information about the amount of sugar present and is called a quantitative test.

Expected results:

- Different food substances can be tested to see which biological molecules are present.

Food	Test for sugar	Test for starch	Test for protein	Test for lipid
Bread	✓	✓	✓	✓
Beef	✗	✗	✓	✓
Nuts	✗	✗	✓	✓

Required practical 4

pH and enzymes

Aim: To investigate the effect of pH on the rate of reaction of amylase enzyme.

REQUIRED PRACTICAL
Investigate the effect of pH on the rate of reaction of amylase enzyme.

Sample method	Considerations, mistakes and errors
1. Put a test tube containing starch solution and a test tube containing amylase into a water-bath at 37°C. 2. After 5 minutes add the amylase solution to the starch. 3. Every 30 seconds take a drop from the mixture and test it for starch using iodine solution. 4. Record how long it takes for the starch to be completely digested. 5. Repeat the experiment at different pH values using different buffer solutions.	• The solutions need to be left in the water-bath for a while to reach the correct temperature before they are mixed. • After mixing, the tube must be kept in the water-bath. • A buffer solution must be used to keep the reaction mixture at a certain fixed pH.
Variables	**Hazards and risks**
• The independent variable is the one deliberately changed – in this case, the pH. • The dependent variable is the one that is measured – in this case, the time taken for the starch to be digested. • The control variables are kept the same – in this case, temperature, concentration and volume of starch and amylase.	• Care must be taken if a Bunsen burner is used to heat the water-bath. • Take care not to spill iodine solution on the skin.

Key points:
• Different enzymes work best at particular pH values.
• The pH value that gives the fastest rate is called the optimum.
• Changes in pH can cause the active site of the enzyme molecule to change shape and change the rate of reaction.

Expected results:
• How long it takes for the starch to be digested will depend on pH.
• The optimum pH will depend on the source of amylase (it could be from fungi or from bacteria).
• To convert to a rate of reaction, calculate:

$$\text{Rate} = \frac{1}{\text{Time}}$$

pH	Time (mins)	Rate
3	5	0.20
5	2	0.50
7	4	0.25
9	10	0.10

Required practical 5

Rate of photosynthesis

Aim: To investigate the effect of light intensity on the rate of photosynthesis using an aquatic organism such as pondweed.

REQUIRED PRACTICAL	
Investigating the effect of light intensity on the rate of photosynthesis.	
Sample method	**Considerations, mistakes and errors**
One of the most common ways of measuring photosynthesis involves observing oxygen output from a piece of pondweed: 1. Place a piece of pondweed in a beaker and shine a light at it using a lamp a specific distance away. 2. Record the number of bubbles of gas coming out of the pondweed in one minute. 3. Repeat this with the lamp at different distances from the pondweed.	• It is best to take at least two readings at each distance and calculate the mean of the number of bubbles. • Carbon dioxide is provided by adding a small amount of sodium hydrogen carbonate to the water.
Variables	**Hazards and risks**
• The independent variable is the light intensity (distance from the light). • The dependent variable is the number of bubbles in one minute. • The control variables are the piece of pondweed, the temperature and the concentration of carbon dioxide.	• Care must be taken to avoid any water being dropped onto the hot lightbulb.

Key points:

• This experiment assumes that the gas bubbles given off by the pondweed are all the same size.
• It is possible to collect the gas and measure the volume but this is a more complicated method.
• It is important to wait for several minutes after moving the light source before taking a reading.

Glass tube
Water
Beaker
Funnel
Pondweed

Expected results:

• As the distance between the light source and the pondweed decreases, the rate of photosynthesis increases until light intensity stops being the limiting factor.
• Light intensity can be found from the distance by using the inverse square law:

Light intensity $= \dfrac{1}{Distance^2}$

Distance in m	Light intensity	Number of bubbles per minute
0.1	100	30
0.2	25	30
0.4	6	21
0.7	2	11
1.0	1	4

Required practical 6

Reaction times

Aim: To plan and carry out an investigation into the effect of a factor on human reaction time.

REQUIRED PRACTICAL	
Investigating the effect of a factor on human reaction times.	
Sample method	**Considerations, mistakes and errors**
Reaction time can be investigated by seeing how quickly a dropped ruler can be caught between finger and thumb: 1. The experimenter holds a metre ruler near the end. 2. The subject has their finger and thumb a small distance apart, either side of the ruler, on the 50 cm line. 3. The experimenter lets go of the ruler and the subject has to trap it. 4. The distance the ruler travels from the 50 cm line is noted. 5. The experiment is repeated on subjects that have just drunk coffee or cola and subjects that have not.	• It is very difficult to control the variables in this experiment. • To obtain reliable results, large numbers of subjects need to be tested and averages taken.
Variables	**Hazards and risks**
• The independent variable is whether the subject has taken in caffeine or not. • The dependent variable is the distance that the ruler travels. • The control variables are the age and sex of the subjects.	• There are limited risks with this experiment.

Key points:

- Any differences in reaction times are likely to be quite small so it is important to repeat this experiment many times and take a mean of the results.
- Coffee and cola drinks usually contain the chemical caffeine.

Expected results:

- In most people, caffeine reduces reaction times and so the distance the ruler travels should decrease in subjects that have drunk coffee or cola.
- It is possible to convert the distance the ruler drops (d) to reaction time using this formula:

Reaction time = $\sqrt{\dfrac{2d}{9.8}}$

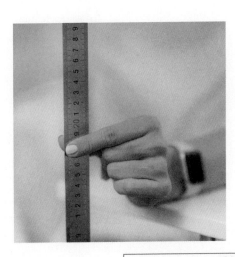

Required practical 7

Ecological sampling

Aim: To measure the population size of a common species in a habitat.

REQUIRED PRACTICAL	
Investigating the population size of a common species in a habitat.	
Sample method	**Considerations, mistakes and errors**
1. Place a quadrat on the ground at random. 2. Count the number of individual plants of one species in the quadrat. 3. Repeat this process a number of times and work out the mean number of plants. 4. Work out the mean number of plants in 1 m². 5. Measure the area of the whole habitat and multiply the number of plants in 1 m² by the whole area.	• The main consideration in the experiment is making sure that the quadrats are placed at random. Using random numbers to act as coordinates can help with this. • The more samples that are taken, then the more accurate the estimate should be.
Variables	**Hazards and risks**
• The independent variable is the number of plants in the quadrat.	• Care should be taken to wash hands after ecology work in a habitat.

Key points:
- In this investigation it is often impractical to count all the organisms in a habitat so a sampling method is used.
- Quadrats are used to sample plants, whereas nets or traps are often used to sample animals.
- Quadrats are often a square with sides 0.5 m. Therefore, they cover an area of $0.5 \times 0.5 = 0.25\,m^2$.

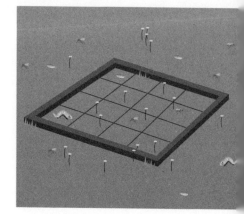

Expected results:
- In an investigation, ten results were taken in a field.
- The field has an area of 100 m².
- In the ten quadrats, there were 10, 6, 7, 5, 9, 8, 7, 7, 8 and 8 plants.
- The mean number of plants per quadrat is therefore $75 \div 10 = 7.5$.
- As each quadrat is 0.25 m², there must be $7.5 \div 0.25 = 30$ plants per m².
- The population size in the whole field would therefore be $100 \times 30 = 3000$ plants.

Required practical 8

Soluble salt preparation

Aim: To prepare a pure, dry sample of a soluble salt.

REQUIRED PRACTICAL
Preparation of a pure, dry sample of a soluble salt from an insoluble oxide or carbonate.

Sample method	Hazards and risks
1. Add the metal oxide or carbonate to a warm solution of acid until no more will react. 2. Filter the excess metal oxide or carbonate to leave a solution of the salt. 3. Gently warm the salt solution so that the water evaporates and crystals of salt are formed.	• Corrosive acid can cause damage to eyes, so eye protection must be used. • Hot equipment can cause burns, so care must be taken when the salt solution is warmed.

Key points:

- A neutralisation reaction takes place between an acid and a base.
- The base (a substance that reacts with acid) is an insoluble oxide or insoluble carbonate.
- The acid is warmed to increase the rate of reaction.
- Pure crystals of the salt can be made by patting the crystals dry with absorbent paper or putting in a drying oven to remove the remainder of the solvent.

Expected results:

- Pure, dry crystals of a soluble salt.

Copper oxide

Sulfuric acid

Add copper(II) oxide to sulfuric acid ➡ Filter to remove any unreacted copper oxide ➡ Evaporate using a water bath or electric heater to leave behind blue crystals of the 'salt' copper(II) sulfate

Required practical 9

Electrolysis

Aim: To investigate the electrolysis of different aqueous solutions with inert electrodes.

REQUIRED PRACTICAL	
Investigate what happens when aqueous solutions are electrolysed using inert electrodes.	
Sample method	**Hazards and risks**
1. Set up the equipment as shown in the diagram. 2. Pass an electric current through the aqueous solution. 3. Observe the products formed at each inert electrode.	• A low voltage must be used to prevent an electric shock. • The room must be well ventilated, and the experiment must only be carried out for a short period of time, to prevent exposure to dangerous levels of gas.

Key points:

- Inert electrodes do not take part in the chemical reaction. They can be made from unreactive metals like platinum but are more usually made from carbon graphite as it is cheap.
- Electrolytes must be ionic compounds where the ions are free to move and carry the charge.
- At the anode (positive electrode):
  - halogens are oxidised if the electrolyte is a halide
  - oxygen is made if the electrolyte does not contain a halide.
- At the cathode (negative electrode):
  - metals below hydrogen in the reactivity series are reduced
  - hydrogen is released if the metal is above hydrogen in the reactivity series.

Expected results:

- Effervescence will be observed at the electrodes if gases are made. These gases can be collected under displacement and tested:
  - hydrogen: lighted splint causes a pop
  - oxygen: relights a glowing splint
  - chlorine: damp litmus paper is bleached.
- Colour change on the cathode as solid metal is deposited if the metal in the electrolyte is below hydrogen in the reactivity series.

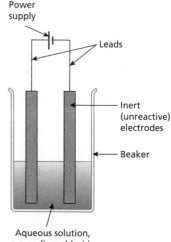

Power supply

Leads

Inert (unreactive) electrodes

Beaker

Aqueous solution, e.g. sodium chloride solution

Required practical 10

Energy transfer

Aim: To determine the energy change that occurs in a solution-based chemical reaction.

<table>
<tr><td colspan="2" align="center">REQUIRED PRACTICAL</td></tr>
<tr><td colspan="2" align="center">Investigate the variables that effect temperature changes in reacting solutions.</td></tr>
<tr>
<td>Sample method</td>
<td>Considerations, mistakes and errors</td>
</tr>
<tr>
<td>

1. Set up the equipment as shown.
2. Take the temperature of the liquid.
3. Add the other reactant, e.g. metal powder and stir.
4. Record the highest temperature that the reaction mixture reaches.
5. Calculate the temperature change for the reaction.

</td>
<td>

- For metal and acid reactions, there should be a correlation between the reactivity of the metal and temperature change, i.e. the more reactive the metal, the greater the temperature change.
- When a measurement is made there is always some uncertainty about the results obtained. For example, if the experiment is repeated three times and temperature changes of 3°C, 4°C and 5°C are recorded:
  - the range of results is from 3°C to 5°C
  - the mean (average) = $\frac{(3 + 4 + 5)}{3}$ = 4°C.

</td>
</tr>
<tr>
<td>Variables</td>
<td>Hazards and risks</td>
</tr>
<tr>
<td>

In the investigation between the reaction of metals and acids:
- The independent variable is the metal used.
- The dependent variable is the temperature change.
- The control variables are the type, concentration and volume of acid.

</td>
<td>

- There is a low risk of a corrosive acid damaging the experimenter's eye, so eye protection must be used.

</td>
</tr>
</table>

Key points:

- The biggest source of error is heat lost to the surroundings.
- The reaction mixture should be stirred to ensure that the temperature is the same throughout the mixture.
- A Pyrex beaker is added to stabilise the equipment and reduce the likelihood that it will fall over.

Expected results:

- A temperature rise indicates that the reaction is exothermic.
- A temperature fall indicates that the reaction is endothermic.
- Exothermic reactions include neutralisation reactions between:
  - acids reacting with metals
  - acids reacting with insoluble bases like carbonates
  - acids reacting with soluble bases or alkalis.
- Displacement of metals also causes a rise in temperature, indicating that the reaction is exothermic.

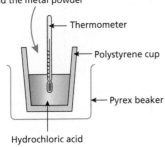

Add the metal powder

Thermometer

Polystyrene cup

Pyrex beaker

Hydrochloric acid

Required practical 11

Rate of reaction

Aim: To investigate how varying the concentration of a solution affect the rates of reactions.

REQUIRED PRACTICAL	
Investigate how changes in concentration affect the rates of reactions by methods involving the production of gas or a colour change.	
This investigation uses the reaction between sodium thiosulfate and hydrochloric acid. **Sample method** 1. Set up the equipment as shown. 2. Add the hydrochloric acid to the flask and swirl to mix the reactants. 3. Start the timer. 4. Watch the cross through the flask. 5. When the cross is no longer visible, stop the timer. 6. Repeat the experiment using hydrochloric acid of a different concentration.	**Considerations, mistakes and errors** • There should be a correlation between the concentration of the acid and the time taken for the cross to 'disappear'. • The higher the concentration of the acid, the faster the rate of reaction, and the shorter the time for the cross to 'disappear'.
Variables • The independent variable is the concentration of the acid. • The dependent variable is the time it takes for the cross to 'disappear'. • The control variables are the volume of acid and the concentration and volume of sodium thiosulfate.	**Hazards and risks** • Corrosive acid can damage eyes, so eye protection must be used. • Sulfur dioxide gas can trigger an asthma attack, so the temperature must always be kept below 50°C.

Key points:
- Rate of reaction can be monitored by the change in a reactant or product.
- If a gas is involved in the reaction, a top pan balance can monitor the mass change in an open system.
- If a gas is made in a reaction, it can be collected by displacement or by a gas syringe and the volume measured.
- If a chemical reaction causes a precipitate to be formed, then turbidity (cloudiness) can be used to monitor the reaction.

Add dilute hydrochloric acid

Timer

Flask

Paper with cross drawn on it

Sodium thiosulfate

$$\text{Gradient} = \frac{\text{difference in the amount of product formed or reactant used}}{\text{time}}$$

Amount of product formed or amount of reactant used up

A

B

Time

The graph shows that reaction A is faster than reaction B.

Expected results:
- The faster the mass changes or gas is collected, the faster the rate of reaction. Continuous data can be collected and a graph drawn. The steeper the gradient, the faster the rate of reaction. So, A would have a higher concentration than B, as shown in the graph.

Required practical 12

Paper chromatography

Aim: To use paper chromatography to generate and interpret a chromatogram.

REQUIRED PRACTICAL	
Investigating how paper chromatography can be used to separate and tell the difference between coloured substances.	
Sample method	**Considerations, mistakes and errors**
1. Draw a 'start line', in pencil, on a piece of absorbent paper. 2. Put samples of five known food colourings (A, B, C, D and E), and the unknown substance (X), on the 'start line'. 3. Dip the paper into a solvent. 4. Wait for the solvent to travel to the top of the paper. 5. Identify substance X by comparing the horizontal spots with the results of A, B, C, D and E.	• Only ever use pencil to draw the start line, as ink will disolve and affect your results.

Key points:

- The solvent front must be marked before the chromatogram is dried as it will not remain visible.
- A lid can be placed on the chromatography tank to slow down the wicking of the solvent and improve the separation.
- If a dye does not move from the start line, it is not soluble in the solvent.
- A pure substance will have only one dot on a chromatogram.
- Dots in different lines at the same height will be the same substance.

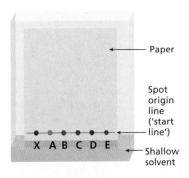

Paper

Spot origin line ('start line')

X A B C D E

Shallow solvent

Expected results:

- A dry chromatogram which shows the separation.
- The R_f values of each spot can be calculated. For example, the blue spot R_f value $= \frac{4}{10} = 0.4$.

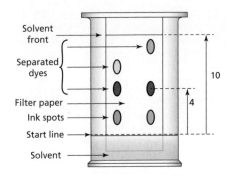

Solvent front

Separated dyes

Filter paper

Ink spots

Start line

Solvent

10

4

Required practical 13

Water

Aim: To analyse and purify water samples.

REQUIRED PRACTICAL	
Analysis and purification of water samples from different sources.	
Sample method	**Hazards and risk**
1. Use a pH probe or suitable indicator to determine the pH of the sample. 2. Set up the equipment as shown. 3. Heat a set volume of water sample to 100°C so that the water changes from liquid to gas. 4. The pure water collects in the condenser and changes state from gas to liquid. Collect this pure water in a beaker. 5. When all the water from the sample has evaporated, measure the mass of solid that remains to find the amount of dissolved solids present in the sample.	• There is a risk of the experimenter burning themselves on hot equipment, so care must be taken during and after the heating process.

Key points:

- pH can be measured using universal indicator and comparing the colour that it turns with the indicator chart.
- pH can also be measured with a pH probe. If the pH probe is connected to a machine to record the data, then it becomes a datalogger.
- The mass of the dissolved solids is measured using a top pan balance.

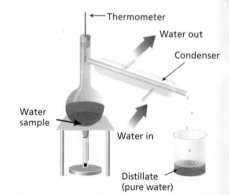

Expected results:

Water sample	pH	Mass of solid obtained by evaporating 50 cm³ of water sample (g)
Distilled	7	0.00
Mineral	9	1.74
Carbonated	4	0.07

- Different samples of water have different masses of residue in them and only pure water will have no residue. As water samples are usually mixtures, their pH varies between different sources.

Required practical 14

Determining specific heat capacity

Aim: To determine the specific heat capacity of a substance.

REQUIRED PRACTICAL	
Investigate the specific heat capacity of materials, linking the decrease of one energy store (or work done) to the increase in temperature and subsequent increase in thermal energy stored.	
Sample method	**Considerations, mistakes and errors**
1. Set up the apparatus as shown. 2. Measure the start temperature. 3. Switch on the electric heater for 1 min. 4. Measure the end temperature. 5. Measure the voltage and current to find the power. 6. Repeat for different liquids. 7. Calculate the specific heat capacity. 8. Compare your results to another group's results. If they get similar answers the experiment is **reproducible**.	• The energy provided by the heater is calculated as power × time. However, it could also be found using a joulemeter. • The specific heat capacity is calculated from the energy provided, the mass of the liquid and the temperature change. • If the temperature rise is too high, energy loss to the surroundings will affect the results.
Variables	**Hazards and risks**
• The independent variable is the type of liquid. • The dependent variable is the temperature. • Control variables are the volume of liquid used and energy provided.	• The electric heater could be very hot so you must not touch it directly. • If the liquids become hot they could boil and spit, so safety goggles must be worn and the heater should not be left on for longer than is necessary. • The liquid can be very hot so you must not touch it.

Key points:

The beaker should be placed on a heat-proof mat to minimise heat loss to the environment. The energy transferred can be calculated using the equation $E = Pt$, where P is the power of the heater, measured in watts (W), and t is the time, measured in seconds (s).
The power of the heater can be determined using the equation $P = IV$, where I is the current, measured in amps (A), and V is the potential difference, measured in volts (V).
A graph of energy transferred against temperature will give a gradient that is equal to mass × specific heat capacity.

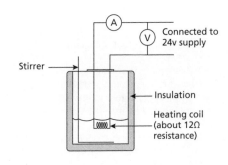

Expected results:

The specific heat capacity of the substance can be compared to data book values. The measured value is likely to be slightly higher due to heat loss to the surroundings.

Required practical 15

Factors affecting resistance

Aim: To investigate the factors that affect the resistance of a piece of wire.

REQUIRED PRACTICAL	
Investigate the factors that affect the resistance of an electrical component.	
Sample method	**Considerations, mistakes and errors**
This example looks at how length affects the resistance of a wire: **1.** Set up the standard test circuit as shown. **2.** Pre-test the circuit and adjust the supply voltage to ensure that there is a measurable difference in readings taken at the shortest and longest lengths. **3.** Use the variable resistor to keep the current through the wire the same at each length. **4.** Record the voltage and current at a range of lengths, using crocodile clips to grip the wire at different points. **5.** Use the voltage and current measurements to calculate the resistance.	• Adjusting the supply voltage to ensure as wide a range of results as possible is important, as measurements could be limited by the precision of the measuring equipment. • The range of measurements to be tested should always include at least five measurements at reasonable intervals. This allows for patterns to be seen without missing what happens in between, but also without taking large numbers of unnecessary measurements.
Variables	**Hazards and risks**

• The independent variable is the length of the wire.
• The dependent variable is the voltage.
• The control variable is the current (which is kept the same, because if it was too high it would cause the wire to get hot and change its resistance).
	• Current flowing through the wire can cause it to get very hot.
	• To avoid being burned by the wire:
	  – a low supply voltage should be used, such as the cell in the diagram
	  – adjust the variable resistor to keep the current low.

Key points:

- The potential difference of the power supply should be kept constant during the investigation.
- The temperature of the wire needs to be maintained at a constant temperature during the investigation. In order to do this, disconnect the wire from the circuit between readings and allow it to cool down.
- As the wire gets shorter, the temperature of the wire will increase quickly, so take the reading as soon as the wire is connected to the circuit and then disconnect it.

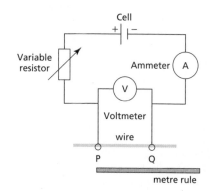

Expected results:

The resistance of the wire should be found to be directly proportional to the resistance, so a graph of resistance against length should produce a straight line that passes through the origin.

Required practical 16

Current–potential difference characteristics

Aim: To investigate the I–V characteristics of a filament lamp, diode and resistor at constant temperature.

REQUIRED PRACTICAL	
Investigate the V–I characteristics of a filament lamp, a diode and a resistor at constant temperature.	
Sample method	**Considerations, mistakes and errors**
1. Set up the standard test circuit as shown.	• Before taking measurements, check the voltage and current with the supply turned off. This will allow zero errors to be identified.
2. Use the variable resistor to adjust the potential difference to the lowest setting across the test component.	• A common error is simply reading the supply voltage as the voltage across the component. At low component resistances, the wires will take a sizeable share of this voltage, resulting in a lower voltage across the component. This is why a voltmeter is used to measure the voltage across the component.
3. Measure the voltage and current for a range of voltage values.	
4. Repeat the experiment at least three times to be able to calculate a mean.	
5. Repeat for the other components to be tested.	
Variables	**Hazards and risks**
• The independent variable is the potential difference across the component (set by the variable resistor) and measured by the voltmeter.	• The main risk is that the filament lamp will get hotter as the current increases and could cause burns. If it overheats, the bulb will 'blow' and must be allowed to cool down before attempting to unscrew and replace it.
• The dependent variable is the current through the component, measured by the ammeter.	

Key points:

The temperature of the component needs to be maintained at a constant temperature during the investigation. In order to do this, disconnect the component from the circuit between readings and allow it to cool down.

In order to take readings for the current flowing in the opposite direction, connect the power supply the other way around.

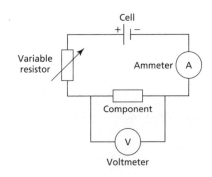

Expected results:

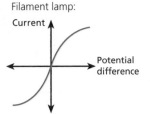

Resistor:

Filament lamp:

Diode:

Required practical 17

Determining densities

Aim: To make and record measurements needed to determine the densities of regular and irregular solid objects and liquids.

REQUIRED PRACTICAL	
Investigate the density of regular solids and irregular solids and liquids.	
Sample method	**Considerations, mistakes and errors**
1. Set the equipment up as shown to determine the density of an irregular solid. 2. Record the height of the water in the measuring cylinder and the mass of the solid being tested. 3. Add the solid being tested to the measuring cylinder. 4. Record the new height of the water in the measuring cylinder. 5. Subtracting the original height from the new height gives the volume of the solid being tested. 6. Now the density of the solid can be calculated.	• If a solid that is less dense than water is tested, the volume measurement will be incorrect because the solid will not be fully submerged. • When reading from the measuring cylinder, the reading should be taken from the bottom of the **meniscus**. • The temperature of the water must be exactly the same throughout all tests, as an increase in temperature could cause the material or water to change volume slightly through expansion.
Variables	**Hazards and risks**
• The independent variable is the material being tested. • The dependent variables are the volume and mass. • The control variable is the temperature of the water.	• There are very few hazards, unless the materials being tested are hazardous or react with water. • The main hazard could be a slip hazard if water is spilt.

Key points:

- When reading the scale on the measuring cylinder, ensure that the measuring cylinder is on a flat surface and that you read the meniscus at eye level.
- Be sure to lower the object into the water carefully so as not to cause splashes that would result in the volume reading less than it should do.

Expected results:

The density of the materials tested can be compared to values from a data book.

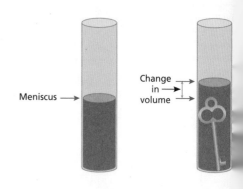

Meniscus

Change in volume

Required practical 18

Force and extension

Aim: To investigate the relationship between force and extension for a spring.

REQUIRED PRACTICAL	
Investigate the relationship between force and extension for a spring.	
Sample method	**Considerations, mistakes and errors**
1. Set up the equipment as shown. 2. Measure the original length of the spring and record this value. 3. Add 100 g (0.98 N) to the mass holder. 4. Measure the extension of the spring and record the result. 5. Repeat steps 2 to 4 for a range of masses from 100 g to 1000 g.	• The extension is the total increase in length from the original unloaded length. It is *not* the total length or the increase each time. • Adding too many masses can stretch the spring too far, which means repeat measurements cannot be made.
Variables	**Hazards and risks**
• The independent variable is the one deliberately changed – in this case, the force on the spring. • The dependent variable is the one that is measured – the extension.	• The biggest hazard in this experiment is masses falling onto the experimenter's feet. To minimise this risk, keep masses to the minimum needed for a good range of results.

Key points:

When reading the scale on the ruler, ensure that you read it at eye level.
A marker, such as a pin mounted in Plasticine, can be used to determine the level of the bottom of the spring.

Expected results:

The force and extension of the spring should be directly proportional, up to the limit of proportionality. A graph of force against extension should give a straight line that passes through the origin.

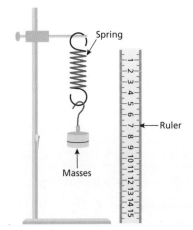

Required practical 19

Acceleration, force and mass

Aim: To investigate the effect of varying the force and/or the mass on the acceleration of an object.

REQUIRED PRACTICAL	
Investigate the effect of varying the force and/or the mass on the acceleration of an object.	
Sample method	**Considerations, mistakes and errors**
1. Set up the equipment as shown with a mass of 100 g. 2. Release the trolley and use light gates or a stopwatch to take the measurements needed to calculate acceleration. 3. Move 100 g (0.98 N) from the trolley onto the mass holder. 4. Repeat steps 2 and 3 until all the masses have been moved from the trolley onto the mass holder. If investigating the mass, keep the force constant by removing a mass from the trolley but not adding it to the holder.	• When changing the force it is important to keep the mass of the system constant. Masses are taken from the trolley to the holder. No extra masses are added. • Fast events often result in timing errors. Repeating results and finding a mean can help reduce the effect of these errors. • If the accelerating force is too low or the mass too high, then frictional effects will cause the results to be inaccurate.
Variables	**Hazards and risks**
• The independent variable is the force or the mass. • The control variable is kept the same. In this case, the force if the mass is changed or the mass if the force is changed.	• The biggest hazard in this experiment is masses falling onto the experimenter's feet. To minimise this risk, masses should be kept to the minimum needed for a good range of results.

Key points:
- Newton's second law states that acceleration is proportional to force, so the greater the force, the greater the acceleration.
- Newton's second law states that acceleration is inversely proportional to mass, so the greater the mass, the smaller the acceleration.

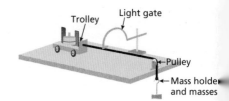

Expected results:

The force and acceleration of the trolley should be directly proportional. A graph of acceleration against force should give a straight line that passes through the origin. The acceleration of the trolley is inversely proportional to the mass, so an acceleration–mass graph will not produce a straight line and so show this inversely proportional relationship.

Required practical 20

Waves in water

Aim: To measure the frequency, wavelength and speed of waves in a ripple tank.

REQUIRED PRACTICAL	
To measure the frequency, wavelength and speed of waves in a ripple tank.	
Sample method	**Considerations, mistakes and errors**
1. Set up the equipment as shown in the diagram. Time how long it takes one wave to travel the length of the tank. Use this to calculate wave speed using: $$speed = \frac{distance}{time}.$$ 2. To find the frequency, count the number of waves passing a fixed point in a second. 3. Estimate the wavelength by using a ruler to measure the peak-to-peak distance as the waves travel. 4. Use a stroboscope to make the same measurements and compare the results. You could also take a picture with a ruler alongside the edge of the tank.	• Using a stroboscope can significantly improve the accuracy of measurements. • By projecting a shadow of the waves onto a screen below the stroboscope, flash speed can be adjusted to make the waves appear stationary. This makes wavelength measurements much more accurate. • For high frequencies that are difficult to count, this equation can be used with the wave speed measurement to calculate the frequency using: $$f = \frac{v}{\lambda}.$$
Variables	**Hazards and risks**
• The key control variable is water depth. It is important to ensure that the depth of the water is kept constant across the tank as, for a given frequency, the depth will affect the speed and wavelength.	• When using a stroboscope there is a risk to people with photo-sensitive epilepsy. It is important to check that there are no at-risk people involved in the experiment or in the area.

Key points:
To determine the wavelength of the wave, you might find it easier to measure the length of 10 waves and divide this value by 10 (this will particularly help if the wavelength is short). The speed of the waves can be determined using the equation:
speed = wavelength × frequency

Expected results:
For a given speed, waves with a longer wavelength will have a lower frequency.

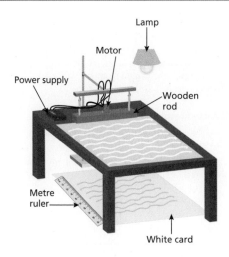

Required Practical 21

Infrared radiation

Aim: To investigate how the amount of infrared radiation absorbed or radiated by a surface depends on the nature of that surface.

REQUIRED PRACTICAL	
Investigate how the amount of infrared radiation absorbed or radiated by a surface depends on the nature of that surface.	
Sample method	**Considerations, mistakes and errors**
1. Take four boiling tubes each painted a different colour: matt black, gloss black, white and silver. 2. Pour hot water into each boiling tube. 3. Measure and record the start temperature of each tube. 4. Measure the temperature of each tube every minute for 10 minutes. 5. The tube that cools fastest emits infrared energy quickest.	• A common error in this experiment is not having the boiling tubes at the same temperature at the start – a hotter tube will cool quicker initially, which can affect results. • Evaporation from the surface of the water can cause cooling too, which will affect the results. To minimise this, block the top of each tube with a bung or a plug of cotton wool.
Variables	**Hazards and risks**
• The independent variable is the colour of the boiling tube. • The dependent variable is the temperature change. • Control variables include volume of water, start temperature and environmental conditions.	• The main hazard is being burned when pouring the hot water and when handling the hot tubes. Using a test tube rack to hold the tubes minimises the need to touch the tubes and means hands can be kept clear when pouring the water into them.

Key points:

• Most of the heat loss from each of the boiling tubes will be due to convection and conduction. The level of conduction and convection will be the same for each boiling tube.
• Any difference in temperature will be due to infrared radiation from the surface.
• Ensure that the same volume of water is added to each of the boiling tubes.

Expected results:

Dark, dull surfaces will be the best emitters of infrared radiation.

Light, shiny surfaces will be the worst emitters of infrared radiation.

Working scientifically

Scientific method

Science is the study of the physical and natural world through **observation** and **experiment**. The **scientific method** is the way that scientists work in a **systematic approach**.

At the start of an investigation, you should set an **aim**. This is what you want to find out in your investigation and is often a question. All of the **data** that you collect in your investigation should help you answer the aim and is called **valid** data.

> There is no need to collect data that doesn't help you answer the aim. For example, if you wanted to know the average height of your classmates you need to collect quantitative data of their heights, but you do not need to collect data on their hair colour as hair colour is not valid data that would help you answer your aim.

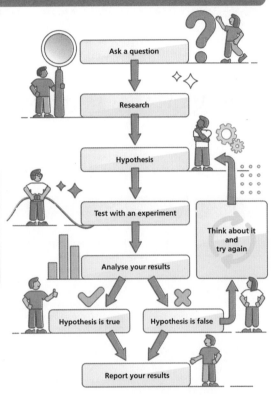

In an **investigation, data** is collected. A researcher will look at the data and think about what it shows. They then try to find a **conclusion**.

A good conclusion:
- describes the relationship between the **independent variable** (the variable you choose to change) and the **dependent variable** (the variable you measure in the experiment)
- is clearly structured
- is explained using scientific knowledge
- links back to the question you want to answer in the investigation, and the **hypothesis** (what you think will happen).

Designing an investigation

In a scientific investigation, the researcher tries to find the relationship between the:
- **independent variable** – the variable they choose to change
- **dependent variable** – the variable that they measure during the investigation.

To make this a **fair test** and give **valid results** (results that answer the aim), other **variables** must be identified and all efforts made to keep them constant in each run of the experiment.

These variables are called **control variables**.

Working scientifically

Risk assessment

Hazards are the damaging effects that are potentially possible from the chemicals and procedures used in an investigation. The **risk** is the likelihood of the damaging effects happening during the investigation. So, a **risk assessment** identifies the hazards, classifies the likelihood of any issues arising and makes suggestions to reduce the potential harm caused. For example, getting acid in the eye could cause damage; by wearing PPE and using small volumes of acid, the risk is mitigated. But if the risk is too high, it may not be possible to safely complete the investigation at all.

Writing a method

A **method**:
- is a step-by-step description of the investigation
- should be written in such a way that anyone could read the method and complete exactly the same investigation.

The method is:
- **reproducible** – if it can be repeated by another person and they obtain the same results
- **repeatable** – if the original experimenter repeats the investigation and obtains the same results.

Data

During an experiment, data is collected and recorded into a results table. This should be drawn in advance so that only the dependent variable results are written down during the investigation.

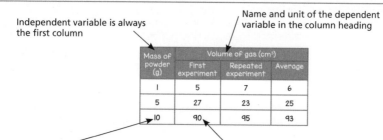

Independent variable is always the first column

Name and unit of the dependent variable in the column heading

Mass of powder (g)	Volume of gas (cm³)		
	First experiment	Repeated experiment	Average
1	5	7	6
5	27	23	25
10	90	95	93

As these values are chosen in the method, this column should be filled in before the investigation starts

Data recorded to the same number of decimal places. Numbers only are needed in the main part of the table as the units are in the column heading

Researchers must keep all the other control variables the same to make the data valid. Valid data is information that can be used to find out if the scientist's idea that is being investigated is correct or incorrect.

In a scientific survey, it is often useful to collect the data in a **frequency table (tally chart)**.

There are different types of data:
- **Qualitative** – using your senses to make observations; this usually includes what you see, what you hear and what you smell.
- **Quantitative** – numbers and values that you have read and recorded from measuring instruments.

Scientists prefer to collect quantitative data because it is:
- **repeatable** and **reproducible** – each time the investigation is repeated, similar results are collected
- **objective** – the data is not influenced by how the person is feeling
- **useful** – graphs and charts can be made which allow **predictions** outside the data collected.

Working scientifically

Results tables

When carrying out investigations, it is important to record data accurately and in a way that is easy to understand.

To make results tables as easy to understand as possible, these conventions should be followed:
- Each column in the table should have a heading, including units if appropriate.
- The data for the independent variable is placed in the first column.
- The data for the dependent variable is placed in the second column.

The purpose of taking repeat readings during investigations is to identify anomalies (data points that sit outside the pattern of the other data points) and to increase the accuracy of the results.

Repeat readings are recorded in further columns in the results table, with a final column showing the mean. Note that when calculating a mean of the data, any anomalous data points are left out.

Accuracy

Scientists want to collect **accurate** data that is close to the true value. But they also want their investigations to produce **reliable** data (data that is similar each time the experiment is repeated).

Imagine if every time you threw a dart at a dartboard you got the bullseye. You would be both reliable and accurate in your throwing.

But if sometimes you hit the bullseye and sometimes you don't, then you are sometimes accurate but you are not reliable.

Precise data have a small range (very little spread) and are similar to the **mean** value. **Random errors** affect precision, but it gives no indication of how close the results are to the true value.

An **accurate result** is close to the accepted value. The accepted value could be:
- written in a text book
- found on a trusted website
- the results collected by your teacher.

The accuracy of the results depends on:
- the quality of the **measuring instruments**
- the skill of the researcher in completing the practical.

Accuracy of results can be improved by:
- carefully following the **method** for the investigation to reduce **random errors**
- correcting any data for systematic errors – remember to calibrate all the measuring equipment
- using an average of the data – take repeated measurements, remove any anomalous results and calculate the **mean**
- selecting a measuring instrument with a **scale** that covers the **range** of values needed for the experiment and is also able to detect the smallest possible change in a measurement.

> Reliable (similar each time you repeat) and accurate (close to the true value) results are the goal of a researcher.

Errors

A **random error** is an error that changes each time the observation is made. This means:
- they are impossible to predict
- they can cause a large range or spread in the data
- they can produce anomalous results.

Random errors can be due to the researcher using the equipment incorrectly. For example, this could include not reading a scale directly in your eye-line.

Working scientifically

A **systematic error** is an error that is the same each time the observation is made. This means:
- a measurement will not be close to the true value
- they are possible to correct in the data.

Systematic errors are often due to the equipment not being **calibrated** correctly. For example, this could include a zero error when you use a balance without first setting it to zero, so it measures a mass as heavier or lighter than it actually is.

Zero errors can be managed by subtracting the initial reading from the final reading. For example, if the balance reads 0.5 g before anything is placed on it, and then 10 g of salt are placed on it, the final reading will be 10.5 g. To determine the true measured value, we would subtract 0.5 g from 10.5 g to give a measured value of 10 g.

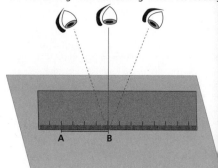

Parallax error

Incorrect reading   Correct reading   Incorrect reading

A   B

Uncertainty is the interval within which the true value is found, with a given level of confidence or probability. Whenever a measurement is made there is always some uncertainty about the result obtained. Estimations can be made about the uncertainty by considering the spread or range of data as well as how a reading compares to the mean.

There will always be errors in an investigation. But, with careful planning of a method and attention to detail during the practical, they can be minimised. As technology improves and we have more accurate measuring equipment, errors will be reduced.

Units

Units are used so that measurements can be compared. There are **quantities** that are often measured in science and you need to know their standard units:

Variable	Unit	Symbol
time	seconds	s
length	metres	m
mass	grams	g
volume	cubic centimetres	cm³
speed	metres per second	m/s
density	grams per cubic centimetre	g/cm³
force	newton	N

Prefixes and standard form

Scientists often work with numbers that are very large or very small.

Prefixes and **standard form** are often used to make numbers more manageable:
- Prefixes are standardised words placed before units used to cut down the numbers that need to be used (see the table on the next page).
- Standard form is a way of writing numbers using the powers of 10.
  So, for example: 1 Gigametre (Gm) = $1\,000\,000\,000$ m = 1×10^9 metres (see the table on the next page).

Be prepared to convert a value with one prefix to a value with a different prefix. You may also be asked to convert a value to standard form.

Working scientifically

Prefix	Multiple size	Standard form	Prefix abbreviation
Tera-	1 000 000 000 000	$\times 10^{12}$	T
Giga-	1 000 000 000	$\times 10^{9}$	G
Mega-	1 000 000	$\times 10^{6}$	M
Kilo-	1000	$\times 10^{3}$	k
Centi-	0.01	$\times 10^{-2}$	c
Milli-	0.001	$\times 10^{-3}$	m
Micro-	0.000001	$\times 10^{-6}$	μ
Nano-	0.000000001	$\times 10^{-9}$	n

SI units

The table lists the seven base SI (Systeme Internationale) units, on which all other units can be based.

Quantity	SI unit
Mass	Kilogram (kg)
Length	Metre (m)
Time	Second (s)
Electric current	Ampere (A)
Temperature	Kelvin (K)
Amount of substance	Mole (mol)
Luminous intensity	Candela (cd)

Time

In investigations, stopwatches or timers are often used to measure time.

Consider a stopwatch that displays 1:30; we read this as 1 minute and 30 seconds, but this is actually two units. So the value needs to be converted either into 90 seconds or 1.5 minutes. (It should never be written as '1.30 minutes', which is actually 78 seconds!)

Bias and ethics

Bias is when:
- one viewpoint is given, often using persuasive language
- evidence is specially selected
- data is specially collected to support a conclusion
- faulty equipment gives inaccurate results.

Bias can lead to misleading conclusions and could cause harm. Scientists reduce the likelihood of biased results and conclusions by checking each other's work before it is published. This is called **peer review** and gives confidence to the public that the information is likely to be true.

Researchers must consider the **ethics** of their experiments. They must consider the risks and benefits and conclude that it is in humanity's best interests to complete the investigation. They should consider the consequences of their research:
- **personally** – the researcher and subjects
- **socially** – groups of people
- **economically** – costs and financial benefits
- **environmentally** – water, air and land.

> Researchers themselves often need to ask other experts like religious leaders, other scientists and the public to decide if the research is suitable to be completed. The research should only take place if the benefits equal or outweigh the risks.

Maths skills

Graphs

A **graph** is a visual representation that looks at the relationship between two quantities. A graph allows us to see a pattern in the data and to make **predictions**. Points that do not fit the pattern are called **anomalous points**.

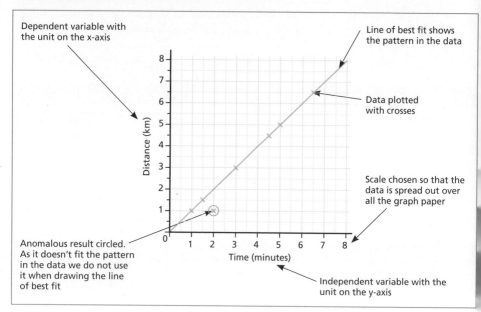

Dependent variable with the unit on the x-axis

Line of best fit shows the pattern in the data

Data plotted with crosses

Scale chosen so that the data is spread out over all the graph paper

Anomalous result circled. As it doesn't fit the pattern in the data we do not use it when drawing the line of best fit

Independent variable with the unit on the y-axis

(Graph: Distance (km) on the vertical axis, Time (minutes) on the horizontal axis)

In science, line of best fit means the line that shows the relationship between the independent variable and the dependent variable. So, in science, lines of best fit can be straight lines, curves or even shapes.

Relationships in numerical data can be described by a **line of best fit** on a graph. **Correlation** happens when changing the independent variable shows a change in the dependent variable. Linear correlation can be shown with a straight line of best fit. The relationship between the independent and dependent variables can be described as **directly proportional** if:
• they produce a straight-line graph, which passes through the origin (0,0)
• the dependent variable doubles when the independent variable has been doubled.

Correlation

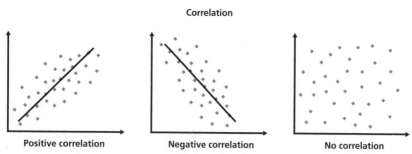

Positive correlation

Negative correlation

No correlation

Maths skills

Range and averages

The **range** of data shows how spread out it is. A small range of data suggests that it is more consistent.

Often, three sets of data are collected for each experiment. If all the values are similar, then the data is reliable. But if a value is not similar you should disregard it as an **anomaly**. There will still be some variation in the results so it is usually worthwhile calculating the **average**.

- **Mean** – the sum of the numbers divided by the amount of numbers
- **Median** – the number in the middle
- **Mode** – the number that appears the most
- **Range** – the difference between the greatest and smallest numbers.

Bar charts	Usually have: • the independent variable on the x-axis, which is either categoric or discrete • the dependent variable on the y-axis, which is a quantity • the height of the bars representing the frequency.	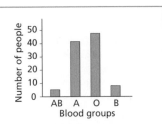
	Always have suitable scales and labels on the axes. Draw the bars accurately using a ruler.	
Pie charts	Usually have: • an independent variable that is categoric or discrete • a dependent variable that can be represented as a ratio or percentage • the angle of each group representing the fraction (out of 360) for that data value • each group labelled either directly on the pie chart or by a colour-coded key.	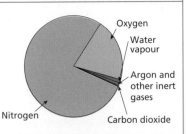
	Use a protractor to work out the fraction each section represents.	
Venn diagrams	A mathematical representation which allows you to compare the features of different groups and consider their relationship to each other. Each group has a circle. Information for each group is placed in their circle and any features shared by more than one group is written in the overlap between their circles.	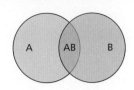

Maths skills

Statistical analysis

Quantities are expressed in numbers. It is useful to express these in digits that suggest that the **magnitude** (size) is **accurate**. Significant figures usually start with the first non-zero number of the quantity and then the number is rounded at that point to show the confidence in number.

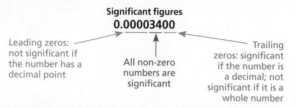

Significant figures
0.00003400

Leading zeros:
not significant if
the number has a
decimal point

All non-zero
numbers are
significant

Trailing
zeros: significant
if the number is
a decimal; not
significant if it is a
whole number

Rearranging equations

During your studies in physics, you will have seen that equations are used to show the relationship between different variables. It is often necessary to rearrange equations in order to calculate an unknown quantity.

The golden rule to follow when rearranging equations is **whatever you do to one side of the equation, you must do to the other**.

Let's take the equation for calculating kinetic energy as an example.

The equation is:

$$E_k = \frac{1}{2}mv^2$$
where
E_k is the kinetic energy, in joules (J)
m is the mass, in kilograms (kg)
v is the speed, in metres per second (m/s)

If the mass is unknown	If the speed is unknown
$E_k = \frac{1}{2}mv^2$	$E_k = \frac{1}{2}mv^2$
Multiply both sides of the equation by 2	Multiply both sides of the equation by 2
$2E_k = mv^2$	$2E_k = mv^2$
Divide both sides of the equation by v^2	Divide both sides of the equation by m
$\frac{2E_k}{v^2} = m$	$\frac{2E_k}{m} = v^2$
	Take the square root of both sides of the equation
	$\sqrt{\frac{2E_k}{m}} = v$

Don't try to carry out more than one step at a time – mistakes can easily be made if you try to complete the rearrangement in one go.

Answers

Biology

Page 9

1. Cell membrane – Controls what enters and leaves the cell
Mitochondria – Respiration
Nucleus – Contains the genetic material
Ribosomes – Protein production

2. a) F **b)** E **c)** D **d)** G **e)** G, E, D

3. There would not be any chloroplasts.

4.

	Mitochondria	Chloroplasts
Are the site of respiration	✓	✗
Contain cell sap	✗	✗
Are surrounded by a double membrane	✓	✓
Contain chlorophyll	✗	✓

5. In mitochondria the membrane is folded inwards. In chloroplasts there are stacks of membranes.

Page 11

1.

	Prokaryotic cells	Eukaryotic cells
Contain mitochondria	✗	✓
Contain DNA	✓	✓
Contain cytoplasm	✓	✓
Contain plasmids	✓	✗

2. In prokaryotic cells they are in the cytoplasm and shaped like a loop. In eukaryotic cells they are in the nucleus and are strands / linear.

3. a) Cell wall: A
Chromosome: C
b) They can pass plasmids between each other.
c) It does not have a flagellum.

4. a) 2 (µm) **b)** 200 times **c)** 400

Page 13

1. The ability to see two objects as separate objects.

2. 0.2 (µm)

3. 1931

4.

	Light microscope	Electron microscope
Produces coloured images	✓	✗
Can be used to view bacteria	✓	✓
Has a resolution greater than 0.0002 mm	✗	✓
Can be used to view living cells	✓	✗

5. a) 5000 **b)** 1.0×10^{-4} **c)** 2

Page 15

1. A length of DNA coding for a protein.

2. gene, chromosome, nucleus, cell

3. One copy is from the mother and one is from the father.

4. The new cell grows / makes sub-cellular structures.

5. So that each cell that is made has the full number of chromosomes.

6. a) 6
b) The chromosomes line up along the centre of the cell, then divide and the copies move to opposite ends of the cell. The cell then divides.

7. a) 10 hours
b) 4.55%
c) Some cells do not live long and so need to be replaced rapidly.

Page 17

1. They have not differentiated / have not specialised.

2. Embryonic stem cells must make all types of cells, but adult stem cells only make certain types of cells.

3. Special areas in plants that contain stem cells.

4. *Any one of:* at the tips of roots; shoots

5. Meristem cells can make all types of cells, but adult stem cells cannot.

6. *Any one of:* diabetes; paralysis

7. Cloning a patient to provide stem cells to cure a disorder.

8. The person's immune system will not attack / reject the cells.

9. All of the plants that are made will have the same genes and so be resistant to the disease.

Page 19

1.

2.

Part of the organism	Substance	Diffuses in	Diffuses out
Human liver cells	Urea		✓
Plant leaves in the light	Carbon dioxide	✓	
Human muscle cells	Glucose	✓	
Fish gills	Oxygen	✓	

3. A change in temperature from 10°C to 20°C

4. Diffusion will be faster on a warmer day.

5. a)

Cube	Surface area in cm²	Volume in cm³	Surface area to volume ratio
A	6	1	6 : 1
B	24	8	3 : 1
C	54	27	2 : 1

b) Organism C because it has the lowest surface area to volume ratio.

Page 21

1. a)

b) The sugar molecules are larger than the water molecules.

2. a) The cylinder would take up water because the cytoplasm is more concentrated than the solution.

b) There would be no net flow of water as the cytoplasm and the solution have the same concentration.

c) The cylinder would lose water because the cytoplasm is less concentrated than the solution.

3. The sugar draws water out of the bacteria and fungi so they cannot reproduce.

4. The movement of substances from a more dilute solution to a more concentrated solution.

5. Sugar in the gut is at a higher concentration than in the bloodstream.

6. Minerals from fertilisers are taken up by active transport. This requires oxygen for respiration. (Active transport requires energy from respiration.)

Page 23

1. embryo, identical, specialised, differentiation, efficient

2.

Cell type	Specialised feature	Function of specialised feature
Sperm	**Has a tail/ flagellum**	Enables the cell to swim to the egg
Muscle	Many mitochondria	**Energy for contraction**
Neurone	**Fatty sheath**	Insulates the axon

3. The sperm cannot fertilise the egg as it cannot digest its way into the egg.

4.

Feature	Xylem cells	Phloem cells	Root hair cells
Walls containing lignin	✓	✗	✗
Holes in the ends of cells	✓	✓	✗
Projection from the side of the cell	✗	✗	✓

5. Minerals are taken up by active transport, which requires energy from respiration. Respiration takes place in the mitochondria.

Page 25

1. muscle – to contract for movement
glandular tissue – to produce substances such as enzymes or hormones
blood – for transport

2. Epithelial – Covers the outside of the stomach
Glandular – Makes digestive juices
Muscle – Churns the food

3. It contains different tissues (connective, endothelial, muscle, elastic).

4. *Any three of:* mouth; oesophagus; stomach; liver; gall bladder; pancreas; large intestine; small intestine; rectum

5. a) xylem and phloem
b) xylem
c) epidermis

Page 27

1. protein, active site, catalysts

2. a) the active site **b)** the substrate

3. Enzymes are specific. Starch cannot fit into the protease active site.

4. 1 The optimum is not the lowest temperature but the temperature that it works best at.

2 As the temperature increases, both molecules will move faster.

3 The higher the temperature the faster the reaction but only up to the optimum.

5.

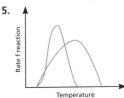

Page 29

1. So that it is small enough to be absorbed into the bloodstream.

2. Carbohydrase – Made in: *any one of:* **Salivary glands**; **Pancreas**
Protease – Made in the stomach and **pancreas**
Lipase – **Lipids; Fatty acids**

3. liver, bile duct, neutralises, stomach, emulsification, lipase

4. They would not be able to store so much bile and so fat digestion would be slower.

Page 31

1. Platelets to **clot blood**; White blood cells to **destroy pathogens**; **Red blood cells** to carry oxygen; Carbon dioxide to be excreted by the **lungs**; **Glucose** for use in respiration

2. a) Lower numbers of red blood cells so the tissues receive less oxygen.

b) Fewer platelets so may have problems with blood clotting.

3.

	Artery	Vein	Capillary
Valves along its length	✗	✓	✗
Wall one cell thick	✗	✗	✓
Takes blood from organs to the heart	✗	✓	✗

Page 33

1. a) D and B

b) It has just passed through the lungs.

c) The blood passes through the heart twice for each circuit of the body.

2. a) Right atrium

b) Right ventricle

c) Valve

d) Aorta

3. a)

b) 2

c) The smaller the distance, the shorter the diffusion pathway for the gases.

Page 35

1. Covid-19; Chickenpox

2. Virus infecting cells in the cervix – Cancer
HIV / AIDS – Tuberculosis
Immune reaction – Asthma

3. It might make them depressed.

4. Skin cancer – UV light
Liver damage – Drinking excess alcohol
Type 2 diabetes – Obesity

5. a) As smoking has decreased, the death rate from lung cancer has decreased.

b) Smoking is decreasing but the death rate is going up. This is because it takes a number of years for smoking to cause death by lung cancer.

Page 37

1. Aorta

2. They supply the heart muscle with oxygen and glucose which is needed for respiration. This provides the energy for the muscle to contract.

3. fatty, oxygen, coronary

4. Blood would leak back into the atrium when the ventricle contracts.

5. a) stents   **b)** statins

6. A malignant tumour can spread around the body.

7. Cancer cells are the body's own cells so a drug that kills cancer cells can damage body cells.

Page 39

1. a) phloem **b)** meristem **c)** xylem

2. It contains many chloroplasts for photosynthesis.

3. upper epidermis, palisade mesophyll, spongy mesophyll, lower epidermis

4. palisade mesophyll, spongy mesophyll, lower epidermis

5. *Any two of:* Leaf B has two layers of palisade cells rather than one; Leaf A has more stomata; Leaf B has a thicker waxy cuticle

6. a) 14 **b)** 0.071 mm² **c)** 197 (*Accept* 198)

Page 41

1. The contents of the root hair cells are more concentrated than the solution surrounding the soil particles.

2. Loss of water from the leaves of a plant.

3. a) Water was taken up by the roots and passed up through the plants to be lost by the leaves.

b) Tube B. Most of the stomata are on the underside of the leaves and are blocked in tube B.

4. Windy and warm

5. The light causes the stomata to be open.

6. Sucrose

Page 43

1. A pathogen causes disease. A vector spreads pathogens but is not affected by them.

2. The tissue will trap the pathogens passed out in the sneeze.

Individuals that have the disease can be isolated to stop them passing on the pathogen.

3.

	Measles	AIDS
Caused by a virus	✓	✓
Spread by breathing in droplets	✓	
Spread by sexual contact		✓
Is usually vaccinated against	✓	
Can lead to death	✓	✓

4. a)

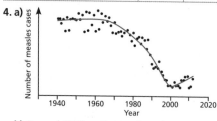

b) Around 1968 as the number of cases started to drop then.

c) The number of cases started to rise as people were worried about having their children vaccinated.

Page 45

1. Poisons that are produced by microorganisms.

2. Clean – This will kill bacteria and prevent them getting onto food.

Separate – This will prevent bacteria on raw food being transferred to cooked foods.

Cook – This will kill any bacteria in the raw food.

Chill – This will slow down the reproduction of the bacteria.

3. protist, vectors, biting

4. a) a fungus

b) using fungicides

c) There is less leaf area for photosynthesis so fewer sugars are made to produce substances needed for growth.

Page 47

1. a) A

b) C

c) Mucus cells release mucus to trap the pathogens. Cilia waft the mucus up to the mouth to be swallowed.

2. Sebum coats the skin and kills pathogens.

3. Antibody – A protein that attaches to the surface of antigens

Antigen – A molecule on the surface of a pathogen

Antitoxin – A molecule that neutralises poisonous substances

4. a) R, P, S, Q

b) The pathogens are being engulfed by white blood cells by phagocytosis.

Page 49

1. white blood cells, antibodies, memory cells, immunity

2. a) There is less of a delay before it starts; Production is quicker; Antibody levels reach a higher peak.

b) There are already memory cells in the blood that are ready to produce antibodies.

3. *Any one of:* antigens; dead pathogens; weakened/inactive pathogens

4. The vaccination might contain weakened but live pathogens, so the person has a mild form of the disease.

Page 51

1. Antibiotics do not affect viruses.
Antibiotics are used in the body not on objects.
MRSA is resistant to, not immune to, antibiotics.

2. The more resistant bacteria might still be alive and would be able to survive and reproduce, producing a more resistant strain.

3. Aspirin – Willow trees
Digitalis – Foxgloves
Penicillin – A fungus

4. New strains of bacteria are appearing with resistance to current antibiotics.

5. a) 16%
   b) *Any one of:* using cells; using tissues
   c) Monkeys are more like humans than mice are, but people have stronger ethical objections about using them.

Page 53

1. sunlight, carbon dioxide, water, glucose, oxygen

2. $6CO_2 + 6H_2O \rightarrow C_6H_{12}O_6 + 6O_2$

3. Reactions that transfer more energy or heat in than they give out.

4. Less chlorophyll in the leaf so less photosynthesis. That results in fewer sugars produced to make chemicals needed for growth.

5. Amino acids – Making proteins
   Fats – Energy storage
   Cellulose – Making cell walls
   Chlorophyll – Trapping sunlight

6. Glucose and starch both contain carbon, hydrogen and oxygen only. Amino acids contain nitrogen as well.

Page 55

1. Light provides the energy for photosynthesis: higher light intensity means more energy.

2. Magnesium ions are needed to make chlorophyll, which is required to trap light.

3. At high temperatures it will not level off but decrease.
   Enzymes cannot die but denature at high temperatures.

4. The heater can provide a higher temperature and the lighting provides more light energy. Temperature and light are less likely to be limiting factors.

5. a) A = light intensity   C = CO_2 concentration
   b) temperature

Page 57

1. An exothermic reaction that releases energy.

2. *Any two of:* movement; keeping warm; chemical reactions to build larger molecules; active transport; the breakdown of excess proteins into urea for excretion

3. Proteins – Amino acids
   Lipids – Fatty acids and glycerol
   Glycogen – Glucose

4. aerobic, carbon dioxide

5. a) The maggots take in oxygen and give out carbon dioxide. The carbon dioxide is absorbed by the solution so the volume of gas in the test tube decreases.
   b) Measure the distance the liquid moves along the scale for a certain time. Then convert the distance to a volume of gas.

6. $C_6H_{12}O_6 + 6O_2 \rightarrow 6CO_2 + 6H_2O$

Page 59

1. In yeast it produces carbon dioxide

2. a) Carbon dioxide given off by the yeast pushes gas through the delivery tube.
   b) The layer of oil.
   c) Count the number of bubbles released in a certain time.

3. $C_6H_{12}O_6 \rightarrow 2C_2H_5OH + 2CO_2$

4. So the muscles can receive more oxygen and glucose for respiration as they need to contract more.

5. a) lactic acid
   b) The muscles are not receiving enough oxygen for aerobic respiration so anaerobic respiration occurs.
   c) It is transported to the liver and converted back to glucose.

Page 61

1. Blood glucose levels: Glucose is the main substance used for respiration.
   Body temperature: Changes to body temperature can decrease enzyme action.

2. Water would enter the blood cells by osmosis, causing them to burst due to lack of a cell wall.

3. Body temperature

4. Coordinator – Processing information
   Effector – Bringing about a response
   Receptor – Detecting a stimulus

5. Effector = Heart muscle
   Receptor = Cells detecting the stretching of blood vessels
   Coordinator = Brain

Page 63

1. a) Central nervous system
   b) Brain and spinal cord

2. a) By diffusion
   b) To allow a nerve impulse in one neurone to produce an impulse in another neurone.

3. It has a similar shape to the neurotransmitter molecule. It will fit into receptor sites at synapses and blocks them, preventing the pain signals from reaching the brain.

4. receptor → sensory neurone → relay neurone → motor neurone → effector
5. Involving the brain would slow it down.
6. It protects the body.
   It happens almost immediately.

Page 65
1. endocrine glands, plasma, target
2. They release the hormones straight into the blood not into ducts.
3.

	Nervous	Hormonal
Type of signal used	**electrical**	chemical
Speed of response	fast	**slow**
Length of response	**short**	**long**
Number of targets affected	**few**	numerous

4. a) pituitary gland
   b) ovaries
   c) testosterone
   d) adrenal
   e) thyroid
5. a) It increases the heart rate and breathing rate, which boosts the delivery of oxygen and glucose to the brain and muscles.
   b) It needs to prepare the body for immediate action and once the danger has passed, the breathing and heart rate need to return to normal or the body might be harmed.

Page 67
1. high blood glucose levels
2. liver and muscles
3. Allows more glucose to enter cells.
   Stimulates glucose to be converted to glycogen.
4.

	Insulin	Glucagon
Produced by the pancreas	✓	✓
Released in response to an increase in blood glucose	✓	✗
Causes more glucose to enter cells	✓	✗
Stimulates the breakdown of glycogen	✗	✓

5. The cells have stopped responding to insulin.
6. a) The person with Type 2 diabetes has a higher level of blood glucose.
   b) In the person with Type 2 diabetes the insulin does not cause such a large decrease in blood glucose level because the cells have stopped responding to insulin.

c) After the injection, the person's blood glucose levels would fall to levels similar to the person without Type 2 diabetes.

Page 69
1. a) E b) D c) A d) A e) E
2. FSH – Stimulating egg development
   LH – Stimulating ovulation
   Oestrogen – Repairing the lining of the uterus
   Progesterone – Maintaining the lining of the uterus
3. a) oestrogen
   b) Stimulates LH release, which will trigger ovulation.
   c) Causes the lining to break down and pass out as menstruation.

Page 71
1.

	Hormonal	Non-hormonal
Oral contraceptive	✓	
Condom		✓
Skin patch	✓	
Diaphragm		✓
Sterilisation		✓

2. Spermicidal cream – Kills sperm
   Condom – Stops sperm entering the vagina after release
   Contraceptive pill – Inhibits egg release
   Female sterilisation – Prevents an egg leaving the oviduct
3. It is very effective at preventing pregnancy and other methods are not required. However, it is not easily reversed and so is fairly permanent.
4. The eggs are removed from the ovaries and the embryo put into the uterus so the blockage is bypassed.
5. The process does not have a high success rate. Further treatments will be easier or cheaper.

Page 73
1. a) It produces runners, which touch the ground and root. The runners then rot away, leaving the new plant growing.
   b) It has flowers.
2. Children receive chromosomes from both parents and so share some of the same alleles as each parent.
3. false, false, true, true
4. A

5.

Type of cell	Number of chromosomes
Elephant skin cell	56
Elephant sperm cell	28
Elephant cheek cell	56
Elephant egg cell	28

Page 75

1. A length of DNA that codes for a protein.

2. gene   chromosome   nucleus   cell

3. a) Q and P

  b) They can reduce the risk factors such as stopping smoking and eating less saturated fat.

4. gene, two, genotype, phenotype

Page 77

1. a) It is controlled by one gene.

  b) It has a lower-case letter.

  c)

		Mother	
		F	f
Father	F	FF	Ff
	f	Ff	ff

    Probability = 25%

2. a) Both Gg

  b) 15 grey and 5 white.

  c) The predicted ratio can be affected by chance and the numbers are quite small.

3. An egg always contains one X chromosome.

Page 79

1. a) E   **b)** B   **c)** G   **d)** B

2. a) A line graph because the data is continuous.

  b) A bar chart because the data is in discrete categories.

3. a) D   **b)** B   **c)** A   **d)** C

4. Tigers and lions are separate species so any offspring produced by a cross will be infertile.

Page 81

1. The remains of organisms from hundreds of thousands of years ago, found in rocks.

2. a) E

  b) The lower layers of rock are formed first and so contain the oldest fossils.

3. a) *Any one of:* From the hard parts of animals that do not decay easily; From parts of organisms that have not decayed because one or more of the conditions needed for decay are absent; When parts of the organisms are replaced by other materials

as they decay; Preserved traces of organisms, e.g. footprints, burrows and root pathways.

  b) They may have been destroyed by geological activity.

4. Bacteria reproduce very quickly.

5. a) mutation

  b) The dark moths were better camouflaged and so less likely to be eaten. They were more likely to survive and mate and pass on the allele for dark.

Page 83

1. To introduce disease resistance in food crops. To produce animals which produce more meat or milk.

2. DNA

3. a) Enzymes cut the gene out of the donor DNA.

  b) Plasmids are used as vectors to insert the gene into the DNA of the recipient.

4. a) He can spray the fields with herbicide to kill all the weeds. That will stop them competing with the crop plants.

  b) The herbicide resistance might spread to other plants, which could become weeds that cannot be killed.

5. Y, V, X, W

Page 85

1. a) Phylum, Order, Genus

  b) *Felis serval*

  c) binomial system

  d) Carl Linnaeus

  e) Jungle cat as it is in the same genus, *Felis*.

  f) Eukaryota

  g) Carl Woese

2. a) chimps / bonobos

  b) 15 million years ago

Page 87

1. Community – All the organisms living in an area
Ecosystem – The living and non-living parts of a habitat
Habitat – An area where organisms live
Population – All the members of a species living in an area

2. a) light, mates, interdependence

  b) stable

3. light, temperature

4. extremophiles

5. *Any one from:* It is very salty, so water will be drawn out of organisms by osmosis; It is very cold, so enzymes will work very slowly

1. a) corn → locusts → lizards → snakes
  b) i) sunlight **ii)** snakes **iii)** locusts
  c) four
  d) two
2. a) predator–prey graph
  b) The peaks for the wolf line come just after the peaks for the moose line.
  The moose numbers are higher than the wolf numbers.

Page 91

1. bacteria and fungi
2. a) A = respiration   B = combustion / burning
  b) P on the arrow from carbon dioxide in the air to plants
3. a) *Answers from top down:* Condensation, Transpiration, Evaporation
  b) Precipitation
4. The water (vapour) condenses to form clouds. This falls as rain, which provides water for other plants to take up through their roots.

Page 93

1. Because species depend on each other for food and water.
2. a) Pond B because more animals were found in the sample.
  b) Pond A because it has a greater variety of different species.
  c) Pond A because it has greatest biodiversity.
3. a) Increased burning of fossil fuels.
  b) Acid rain
4. *Any one suitable answer, e.g.* Greater demand for energy for heating; More cars

Page 95

1. carbon dioxide, methane
2. Sun, warms / hits, reflected / re-radiated, greenhouse, increase, global
3. a) To give a true value as there are few local sources of carbon dioxide.
  b) More carbon dioxide is taken in by plants in the summer as there is more photosynthesis.
4. Malaria could start occurring in non-tropical countries as the mosquitos could survive there.
5. Animal populations may decrease as there is less food for them.

Page 97

1. a) Carbon dioxide is taken in during photosynthesis and the carbon is stored in wood.

b) *Any two of:* to use the land to grow crops; to keep animals; to grow crops to make biofuel; to build houses/roads
2. a) They take in more carbon dioxide than they give out methane, so reduce greenhouse gases in the air.
  b) When they are drained, they give out carbon dioxide and so increase greenhouse gases in the air.
3. Restoring hedgerows – Lack of nesting sites for birds
  Recycling rubbish – Large areas of land used for landfill
  Breeding programmes – Populations of species decreasing to low numbers
4. The carbon dioxide is now being taken in again rather than given out but methane output has increased. Therefore, there is still a net output of greenhouse gases.

Chemistry

Page 99

1. The smallest particle of an element that can exist on its own.
2. All the atoms (particles) are the same.
3. (Just under) 100
4. neutron; proton; electron
5. A charged atom
6. a) Positive   **b)** Negative
7. Metals
8. They are different forms of the same element.
  Or: They are atoms with the same atomic (proton) number but a different mass number.
9. a) They have the same number of protons in the nucleus.
  Or: They have the same number and arrangement of electrons in electron shells.
  b) They have different numbers of neutrons in the nucleus.
10. 3

Page 101

1. A chemical reaction
2. a) 3   **b)** 2
3. Chemical reactions
4. More than one substance not chemically joined together.
5. A mixture that is designed to be a useful product.

6. Tin (Sn) and Copper (Cu)

7. **a)** Insoluble solid from a liquid / solution
   b) Distillation
   c) Similarity: Both techniques separate solutions.
   Difference: Distillation collects the solvent whereas crystallisation collects the solute.
   d) Fractional distillation

Page 103

1. To understand observations and make predictions.

2. New evidence (data) is collected.

3. Solid spheres that could not be divided.

4. Plum pudding model

5. Most of an atom is empty space.

6. A small positive nucleus is at the centre of the atom.

7. James Chadwick

8. nucleus; radius; protons/neutrons; neutrons/protons; positive/+1; neutral/0; 1

Page 105

1. **a)** Atomic mass
   b) Atomic (proton) number

2. **a)** Sodium **b)** 11 **c)** 11 **d)** 12

3. Number of protons and number of electrons

4. Mass number – atomic (proton) number

5. Atomic (proton) number and so the number of protons and electrons

6. The periodic table shows relative atomic mass, not atomic mass.

7. The number of protons in an atom of an element.

Page 107

1. **a)** 2 **b)** 8 **c)** 8

2. From the centre and each energy level is completed before the next one is started.

3. 1

4. 3

5.

2,1

6. Argon

7. A charged atom

8. **a)** 1+
   b)

Sodium ion
2,8

9. All ions have a complete outer shell of electrons. A sodium atom loses one electron and a fluorine atom gains one electron, resulting in both ions having the same electronic structure.

Page 109

1. Putting objects into groups with similar properties.

2. To describe objects without confusion, to see connections between different objects and to make predictions about new objects that are found and fit into a particular group.

3. Not all elements had been discovered and some elements were put in inappropriate groups.

4. **a)** Broadly by increasing atomic weight and by moving some elements around to group them by properties.
   b) He believed that more elements were still to be discovered.
   c) The discovery of elements that matched his predictions, and the discovery of subatomic particles and isotopes.

5. By increasing atomic (proton) number

6. Group 0 (noble gases)

7. solid *(Accept metals)*

Page 111

1. In the first column

2. Alkali metals

3. They all have one electron in the outer shell.

4. 1+

5. **a)** It decreases.
   b) There is a decrease in the force of attraction between the atoms.
   c) It decreases.
   d) It increases.

6. Rubidium oxide (Rb_2O)

7. Burns quickly with a bright yellow flame and a white solid is produced.

8. An observation where you hear fizzing and see bubbles.

Page 113

1. 7th main column of the periodic table / 17th column of the periodic table

2. Halogens

3. All have seven electrons in the outer shell.

4. 1–

5. **a)** It increases.
   b) The molecules get larger and there is a stronger attraction between them.

c) They get darker.

d) It decreases.

6. A 1- ion resulting from a halogen/a halogen atom gaining one electron in the outer shell.

7. A metal compound that contains a halide ion.

8. A reaction in which a more reactive element/halogen takes the place of a less reactive halogen/element from its compound.

9. chlorine + sodium bromide → sodium chloride + bromine

Page 115

1. Last column

2. Noble gases

3. All have a full outer shell of electrons / stable electron configuration.

4. Gases

5. It increases.

6. As relative atomic mass increases, so does the boiling point.

7. **a)** 2   **b)** 8

8. They have a full outer shell of electrons / they are electronically stable

9. **a)** Do not easily form molecules

b) Do not easily form ions

Page 117

1. Non-metals

2. A shared pair of electrons

3. Noble gases/Group 0

4.

5. 3

6. H–N–H
    |
    H

7. A giant structure made of very large molecules, made of smaller repeating units.

8. The atoms are too far apart from each other, and the electrons are not visualised.

9. Dot and cross diagram

Page 119

1. They lose outer shell electrons.

2. 2+

3. Na^+

4. They gain electrons into their outer shell.

5. 1^-

6. O^{2-}

7. **a)** The electrostatic force of attraction between oppositely charged ions

b) A metal and a non-metal

c) From the metal to the non-metal

8. A giant structure / lattice

9. Na_2O

Page 121

1. Substances that contain atoms of only one metallic element.

2. **a)** metallic

b) giant structure

c) in a regular pattern

3. The sharing of the delocalised electrons between the metallic ions in a giant structure.

4. Conductor of heat – Delocalised electrons can transfer the energy.
Malleable (bends and shapes easily) – The layers of atoms easily slide over each other.
High melting and boiling points – Many strong metallic bonds must be broken.
Conductor of electricity – Delocalised electrons are free to move and carry charge.

5. A mixture/formulation of a metal and at least one other element.

6. Steel

Page 123

1. Shared pairs of electrons / covalent bonds

2. Lattices

3. Pure metal elements and alloys

4. Liquid and Aqueous solution

5. All of the atoms in these structures are linked to other atoms by strong bonds. A lot of energy is needed to break any bond.

6. Non-metal elements

7. The greater the size, the greater the melting and boiling points.

8. Electrical insulator – No charged particles are free to move and carry a charge.
Low melting and boiling points – Only relatively weak intermolecular forces of attraction need to be overcome and no bonds are broken.
Soft and brittle – Weak intermolecular forces of attraction are easily broken.

Page 125

1. liquid; gas; solid OR gas; liquid; solid

2. Particles are fixed and are not able to flow past each other.

3. Liquids and solids

4. To explain observations about matter and make predictions about how matter will behave if we change the conditions.

5. Buckminsterfullerene

6. No new substance is made.

7. Melting

8. A substance changes from a gas to a liquid state.

9. Gas

10. Dissolved in water / aqueous solution / solution where water is a solvent.

11. (l)

Page 127

1. 4

2. a) Covalent   **b)** Covalent

3. a) 4
   b) Many strong covalent bonds must be broken to pull apart the atoms from the giant covalent structure.

4. a) The outer shell of each carbon atom.
   b) 3
   c) The layers / planes of atoms slide easily over each other.

5. a) One layer / plane of hexagonal rings of carbon atoms / graphite.
   b) Delocalised electrons are free to move and carry the charge.

6. C_{60}

7. Very high length to diameter ratios (*Accept* high melting and boiling points).

Page 129

1. On the periodic table

2. 12

3. The sum of the atomic masses for each atom in the molecule.

4. a) 2   **b)** 48

5. The sum of the atomic masses for each atom in the molecule.

6 a) 34   **b)** 94

7. The sum of the atomic masses for each atom in the empirical formula.

8. a) 72.7%   **b)** 11%   **c)** 89%

Page 131

1. A new substance is made.

2. No new substance is made.

3. a) Reactants   **b)** Products

4. Goes to

5. Conservation of mass / no atoms are created or destroyed, only rearranged.

6. $2Mg + O_2 \rightarrow 2MgO$

7. The ions involved directly with the chemical reaction.

8. An ion that remains in solution and is unchanged from the reactants to the products.

9. Either oxidation or reduction.

10. $Cu \rightarrow Cu^{2+} + 2e^-$

Page 133

1. Oxygen is added to the fuel.

2. Electrons are lost.

3. Oxygen is removed from the iron oxide.

4. Electrons are gained.

5. Base and acid

6. Salt

7. Electrolysis and thermal

8. An ionic compound that is molten or in aqueous solution.

9. An insoluble solid

10. Filtering

11. A more reactive halogen will take the place of a less reactive halogen in a compound.

12. A more reactive metal will take the place of a less reactive metal in a compound.

Page 135

1.

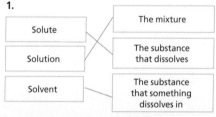

2. A lot of solute in a given volume of solvent.

3. Very little solute in a given volume of solvent.

4. a) 500 cm³   **b)** 1.5 dm³

5. Concentration = mass ÷ volume (*Accept* $C = \frac{m}{V}$)

6. $\frac{10}{2.5}$ = 4 g/dm³

7. Top pan balance

8. *Any one from:* The reaction is reversible; During each transfer, some product is lost; The separation of the product from the reaction mixture leaves some product behind; Other reactions take place.

Page 137

1.

Mole	→	Number of particles in a mole, 6.02×10^{23}
Molar mass	→	The mass of one mole of a substance which is equal to the M_r value in grams
Avogadro constant	→	The chemical measure of the amount of substance

2. Number of moles = mass ÷ relative atomic mass

3. 2 moles

4. Number of moles = mass ÷ relative formula mass

5. 2 moles

6. 80 g

7. The reactant that is fully used up in the chemical reaction.

8. Add too much / so much that the reactant is still visible.

Page 139

1. The outer shell electrons are lost.

2. $Na \rightarrow Na^+ + e^-$

3. A list of metals from most to least reactive.

4. Reaction of metals with water and reaction of metals with acids.

5. Carbon and hydrogen

6. *Any suitable example, e.g.* lead, tin, iron and zinc

7. A more reactive metal will take the place of a less reactive metal in its compound.

8. Magnesium sulfate and zinc

9. Magnesium is more reactive than zinc and magnesium is already in the compound.

10. copper + magnesium sulfate has no reaction, so no word equation.

11. magnesium + copper sulfate → copper + magnesium sulfate

Page 141

1. Uncombined

2. A compound containing a metal that is found in nature.

3. A mineral where it is economical to extract the metal.

4. Reduction

5. When the metal is below carbon in the reactivity series and the metal doesn't react with carbon.

6. iron oxide + carbon → carbon dioxide and iron

7. Global warming and climate change

8. When the metal is above carbon in the reactivity series or reacts with carbon.

9. A lot of energy is needed to melt the metal-containing compound and a lot of electricity is needed.

10. aluminium oxide → aluminium + oxygen

11. The oxygen that is made at the anode immediately reacts with the carbon of the anode and burns it away.

Page 143

1. Oxygen is lost from a substance.

2. Electrons are gained by a substance.

3. Aluminium oxide loses oxygen.

4. Hydrogen ions

5. Oxygen is gained.

6. Electrons are lost.

7. Oxygen is gained by the carbon.

8. A chemical change where both oxidation and reduction happen at the same time.

9. As one reactant is oxidised, the other reactant is reduced. So, both oxidation and reduction happen at the same time.

10. a) Magnesium
  b) copper(II) ions / copper(II) sulfate
  c) Both oxidation and reduction are happening at the same time.

Page 145

1. $H^+(aq)$

2. Sulfuric acid

3. Metals above hydrogen in the reactivity series.

4. Neutralisation

5. Calcium sulfate and water

6. Sodium chloride and water

7. potassium hydroxide + sulfuric acid → water + potassium sulfate

8. Carbon dioxide

9. sodium carbonate + nitric acid → sodium nitrate + carbon dioxide + water

10. An ionic compound where the hydrogen in an acid has been swapped out for a metal ion or ammonium ion.

11. Precipitation

Page 147

1. A measure of the acidity / alkalinity of a solution.

2. 0–14

3. Add water to dilute, or add a chemical to neutralise some of the acid or alkali.

4.

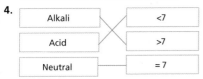

Alkali	<7
Acid	>7
Neutral	= 7

5. Universal indicator

6. pH probe

7. a) Acid
  b) Alkali
  c) Water

Page 149

1. The process of making ions.

2. The greater the amount of $H^+(aq)$, the lower the pH value.

3. a) An acid (substance that releases $H^+(aq)$) that fully ionises in solution.
  b) *Any one of:* nitric acid; HNO_3; sulfuric acid; H_2SO_4; hydrochloric acid; HCl

4. $HA \rightarrow H^+ + A^-$

5. $HCl \rightarrow H^+ + Cl^-$

6. a) An acid (substance that releases $H^+(aq)$) that partially ionises in solution.
  b) *Any one of:* citric acid; ethanoic acid (CH_3COOH); carbonic acid (H_2CO_3); any organic acid

7. $HA \rightleftharpoons H^+ + A^-$

8. $CH_3COOH \rightarrow H^+ + CH_3COO^-$
  (*Also accept* $\rightleftharpoons$ *in equation*)

9. $HCOOH \rightarrow H^+ + HCOO^-$
  (*Also accept* $\rightleftharpoons$ *in equation*)

Page 151

1. a) A new substance is made.
  b) Energy is taken in by the system.

2. *Any one of:* positive ions; metal ions; hydrogen ions

3. Electrons are lost from the ions (oxidation) to form neutral atoms.

4. Direct current (dc)

5. Molten (l) or aqueous solution (aq)

6. Carbon (C)

7. *Accept any metals above carbon in the reactivity series.*

8. Cathode

9. decreases; increases

Page 153

1. anode; halogen; oxygen

2. $2Br^- \rightarrow Br_2 + 2e^-$

3. Pure metal

4. Hydrogen

5. Copper

6. $Cu^{2+} + 2e^- \rightarrow Cu$

7. a) Pb **b)** Br_2

8. Carbon in the anode reacts with the oxygen that is produced in the electrolysis and burns the anode away.

Page 155

1. It is the same / constant.

2. Energy is released from the chemical system into the surroundings.

3. Stored chemical energy of the products is less than the reactants.

4. Thermometer

5. a) A diagram which models the energy changes from the start, during and at the end of the reaction.
  b) The difference between the energy level of the reactants and the products.
  c) Reactants have a higher energy level than the products.

6. Hand warmers and self-heating cans

7. H_2O

8. Aqueous solution

Page 157

1. Energy comes from the surroundings into the chemical system.

2. Stored chemical energy in the products is greater than the stored chemical energy in the reactants.

3. Temperature decreases.

4. a) Time / progress of the reaction
  b) Energy

5. The difference in the energy level of the reactant and the product.

6. Thermal decomposition

7. *Any one of:* sherbet in sweets; sports injury packs

8. Reactants have less energy than products / energy is absorbed from the surroundings / temperature decreases.

Page 159

1. a) Chemical bonds break.
  b) Chemical bonds are made.

2. endothermic; exothermic; released

3. $C=O$ is stronger than $C-H$

4. kJ/mol

5. **a)** The sum of the product bonds is stronger than the sum of the reactant bonds.
  b) The sum of the reactant bonds is stronger than the sum of the product bonds.

6. Overall energy change for a reaction = energy needed to break the reactant bonds – energy released when product bonds are made

7. **a)** exothermic reaction
  b) endothermic reaction

Page 161

1. The speed of the chemical reaction / the change in the reactant used or the product formed in a given time.

2. Top pan balance

3. **a)** Gas from the air is a reactant.
  b) Product is a gas and is released to the air.

4. Cloudy

5. Use the disappearing cross method.

6. g/s and cm³/s

7. Time

8. The steeper the gradient, the faster the rate of reaction.

9. The reaction has stopped.

Page 163

1. A scientific model to explain and predict the rate of a chemical reaction.

2. A collision between reactant particles, with energy equal to or greater than activation energy and in the correct orientation.

3. increases; increases

4. There are more collisions between reactants in a given time, but the same percentage are successful, so overall there are more successful collisions in a given time and therefore a faster rate of reaction.

5. It increases.

6. Enzyme

7. They stay the same.

8. B

Page 165

1. A chemical change where the reactants make the products and the products make the reactants.

2. **a)** Heat **b)** Cooling **c)** 176 kJ/mol

3. The forward reaction of a reversible reaction is exothermic. – The backward reaction is endothermic.
The forward reaction of a reversible reaction is endothermic. – The backward reaction is exothermic.

4. Colour change of the powder from blue to white.

5. **a)** A reversible reaction and closed system.
  b) They are the same.
  c) They remain constant.

6. Equilibrium is reached quicker / Increases rate of forward and reverse reaction by the same amount.

Page 167

1. To predict the effect on position of equilibrium / yield of product when conditions of an equilibrium reaction are changed.

2. For a reversible reaction at equilibrium, the system will oppose any change to the conditions.

3. decreases; increases; decreases; increases

4. **a)** It shifts to favour endothermic reaction / increases rate of endothermic reaction.
  b) It shifts to favour exothermic reaction / increases rate of exothermic reaction.

5. **a)** It shifts to the side with the lowest amount of gas.
  b) It shifts to the side with the highest amount of gas.
  c) No effect.

Page 169

1. **a)** Ancient biomass / plankton
  b) In rocks

2. A mixture of hydrocarbons with similar boiling points.

3. Evaporation and condensation

4. Hydrogen; Carbon

5. The larger the molecule, the less flammable it is.

6. The fuel is fully oxidised.

7. Single covalent bonds

8. hydrocarbon + oxygen → carbon dioxide + water

Page 171

1. The endothermic / thermal decomposition of long-chain hydrocarbons to make shorter, more useful, hydrocarbons.

2. Alkanes and alkenes / shorter chain hydrocarbons

3. It is vapourised, passed over a catalyst and undergoes thermal decomposition.

4. a) Catalyst (zeolite / aluminium oxide / silicon dioxide) and temperatures of 550°C.

b) Petrol

5. a) They only contain hydrogen and carbon.

b) They contain two fewer hydrogen atoms than the alkane with the same number of carbon atoms. They contain double bonds.

c) C = C, carbon carbon double bond

6. Bromine water

7. C_3H_6

8. a) Oxygen, carbon dioxide, water

b) Oxygen, carbon / soot, carbon dioxide, carbon monoxide

Page 173

1. Nothing added / in its natural state.

2. Only one substance.

3. Water molecules only.

4. It contains more than one substance and is therefore a mixture.

5. A single type of element or a single type of compound.

6. a) A mixture designed for a specific purpose.

b) It contains more than one substance.

7. *Any one of:* fuels; cleaning agents; paints; medicines; alloys; foods; fertilisers

8. It contains more than one substance.

9. For a particular function / purpose in the end product.

Page 175

1. a)-b)

Stationary phase
Mobile phase

2. The different parts of the mixture are attracted to the paper and solvent by different amounts and this causes them to separate.

3. The results of a chromatography experiment.

4. pure; impure; mixture

5. Retention factor / the ratio of the distance moved by a compound (centre of spot from origin) to the distance moved by the solvent.

6. The line that marks where the solvent travelled to in the paper / stationary phase.

7. a) $R_f \text{ value} = \dfrac{\text{distance travelled by substance}}{\text{distance travelled by solvent}}$

b) The identity of a substance.

Page 177

1. Seeing bubbles and hearing fizzing

2. One of the products is a gas and is lost to the atmosphere.

3. Displacement of water – When the gas has low solubility in water; Downward delivery – When the gas is denser than air; Upward delivery – When the gas is less dense than air

4. a) Hydrogen

b) Carbon dioxide (*Accept* chlorine)

c) Hydrogen

5. Lighted splint and hear a squeaky pop.

6. It relights.

7. Limewater

8. Litmus

9. It decolourises / bleaches / turns white.

Page 179

1. The envelope of gas around our planet.

2. a) Nitrogen **b)** 20% **c)** Four fifths

3. Mars and Venus

4. Volcanic activity

5. Algae

6. Photosynthesis

7. Sedimentary

8. *Any three from:* dissolved into oceans; formed sedimentary rocks; formed fossil fuels; used in photosynthesis

Page 181

1. Water vapour, carbon dioxide and methane

2. To keep global temperatures high enough to support life.

3. They are increasing the proportion of gases, through combustion of fossil fuels, animal farming, rice paddies and landfill.

4. The increase in average world temperatures.

5. The long-term change in weather patterns and temperatures.

6. Increase in world temperatures cause the ice caps to melt and therefore sea level rises.

7. The total amount of greenhouse gases over the full life cycle of a product, service or event.

8. Carbon dioxide equivalent/CO_2e

Page 183

1. Sulfur

2. From engines with high temperature and pressure.

3. a) CO

b) Incomplete combustion
c) It reduces the oxygen-carrying capacity of the blood.
4. Sulfuric acid
5. Nitric acid
6. Rain with a pH lower than natural rain (5.5).
7. Carbon (soot) / solid particles
8. particles; reflect

Page 185
1. To provide warmth, shelter, food and transport.
2. **a)** A resource that is used chemically unchanged to support life and meet people's needs.
   b) A resource that has been chemically changed to make a new material.
3. **a)** Natural resource
   b) Synthetic resource
4. A resource that can be replaced as it is being used.
5. A resource that is being used up faster than the Earth can replace it.
6. Biomass – Renewable; Nuclear – Finite; Coal – Finite; Geothermal – Renewable; Wind – Renewable; Oil – Finite
7. Ensures the needs of the people are met today, whilst ensuring that there are enough resources for future generations too.

Page 187
1. Pure water is water that only contains water molecules. Potable water is any water that is safe to drink (including water that has safe, dissolved substances in).
2. Naturally occurring non-salty water, e.g. rain water. It can be found in rivers, ground water and lakes.
3. Filtering and sterilising
4. The water is passed through filter beds.
5. Ozone or UV light
6. When fresh water supplies are limited.
7. A lot of energy is used.
8. Reverse osmosis and distillation

Page 189
1. To prevent pollution.
2. Organic matter and microbes
3. Harmful chemicals and organic matter
4. Organic matter and harmful microbes
5. Large, insoluble particles
6. **a)** The liquid part of sewage
   b) The solid part of sewage

c) By sedimentation
7. To separate out the sewage sludge (solid) from the liquid (effluent) as they need to be treated differently.
8. Anaerobic bacteria
9. Aerobic bacteria

Page 191
1. Finite natural resource
2. Rocks that have a low percentage of metal in them.
3. *Any one of:* cooking pans; water pipes; electrical wires
4. **a)** Phytomining
   b) Bioleaching
5. A solution that contains metal ions.
6. **a)** *Any one of:* it preserves finite metal ore reserves; it can be used to clean up contaminated soils.
   b) It is time-consuming.
7. A more reactive metal takes the place of a less reactive metal in its compound.
8. **a)** They are reduced. (*Accept* gain electrons or become neutral atoms)
   b) They are reduced. (*Accept* gain electrons or become neutral atoms)

Page 193
1. The environmental impact of a product through all of its stages of manufacture, use and disposal.
2. **a)** Value judgements, numerical values of pollution effects
   b) Use of water, resources, energy sources and production of some wastes
   c) Facts without opinion / bias.
3. It is recycled or put in landfill.
4. A considered decision with sensible conclusions.
5. It must be cleaned and sterilised.
6. It is melted and re-cast / reformed.
7. It is crushed and melted to form new glass products.

Physics
Page 195
1. Energy cannot be created or destroyed, but can be transferred from one store to another.
2. **a)** The pear's gravitational potential energy store
   b) The pear's kinetic energy store
3. **a)** The bicycle's kinetic energy store
   b) Thermal energy store of the brakes
4. *From the left:* chemical; circuit; electric current

5. The kinetic energy of the ball reduces to zero as it is transferred to the thermal energy of the ball and the wall. There is also an energy transfer by waves, as a sound can be heard when the ball makes contact with the wall.

Page 197

1. mass of the object; speed of the object

2. The bus will have greater kinetic energy. This is because they are both moving at the same speed, but the bus has a greater mass. As kinetic energy depends on mass and speed, if the speed is equal, the object with the greater mass will have greater kinetic energy.

3. $E_k = \frac{1}{2} \times 100 \times 7^2 = 2450\,J$ (joules)

4. $E_k = \frac{1}{2} \times 30\,000 \times 56^2$
$E_k = 47\,040\,000\,J = 47\,040\,kJ$

5. $\frac{2E_k}{v^2} = m$
$m = \frac{(2 \times 640)}{4^2} = 80\,kg$

6. $\sqrt{\frac{2E_k}{m}} = v$
$v = \sqrt{\frac{(2 \times 5\,750\,000\,000)}{184\,000}} = 250\,m/s$

Page 199

1. the spring constant; the extension of the spring

2. $E_e = \frac{1}{2} \times 300 \times 0.05^2 = 0.375\,J$

3. Extension = 25 cm − 5 cm = 20 cm
Extension = 0.2 m
$E_e = \frac{1}{2} \times 500 \times 0.2^2 = 10\,J$

4. $20 = \frac{1}{2} \times k \times 0.40^2$
$k = 250\,N/m$

5. mass; gravitational field strength; height

6. $E_p = 0.75 \times 10 \times 1.2 = 9\,J$

7. $E_p = 0.045 \times 10 \times 20 = 9\,J$

8. 637.5 kJ = 637 500 J
$637\,500 = 75 \times 10 \times h$
$h = 850\,m$

Page 201

1. The energy required to raise the temperature of 1 kg of a substance by 1°C.

2. mass; specific heat capacity; change in temperature

3. $\Delta E = 0.0035 \times 385 \times 30$
$\Delta E = 40.4\,J$ (to 3 s.f.)

4. $\Delta E = 2 \times 4200 \times 82 = 688\,800\,J$
$\Delta E = 688.8\,kJ$

5. $\frac{\Delta E}{(c\Delta\theta)} = m$
$m = \frac{78\,750}{(900 \times 35)} = 2.5\,kg$

6. $\frac{\Delta E}{(mc)} = \Delta\theta$
$\Delta\theta = \frac{28\,350}{(1.5 \times 420)} = 45°C$

Page 203

1. joule; second; watt

2. Motor A is more powerful as it transfers the same energy as Motor B in a shorter time.

3. a) Power = $\frac{10\,000\,000}{(20 \times 60)} = 8333\,W$
b) $\frac{8333}{1000} = 8.3\,kW$

4. a) The kettle transfers 2500 joules of energy every second.
b) Power = $\frac{energy}{time}$ so energy = power × time
Energy = 2500 × (3 × 60)
Energy = 450 000 J = 450 kJ

5. a) 2 MW = $2 \times 10^6\,W$
b) Energy = $(2 \times 10^6) \times (30 \times 60)$
Energy = 3 600 000 000 J = 3600 MJ

Page 205

1. Energy cannot be created or destroyed. It can only be transferred from one store to another.

2. Useful – *Any one of:* electrical to light; electrical to sound
Wasted – electrical to thermal

3. Oiling reduces the friction between two surfaces, which reduces the unwanted energy transfer into the thermal store.

4. *Any two of:* cavity wall insulation; loft insulation; double-glazing; draught excluders; curtains; carpets; turn off appliances when they are not in use

5. It is made of a material with a low thermal conductivity to minimise the rate of thermal energy transfer through the material. This makes it effective as an insulator.

Page 207

1. Efficiency is how much of the energy transferred by a device is transferred as a useful output. The higher the efficiency, the greater the useful energy transfer.

2. An efficiency of greater than 1 would mean that the device was usefully transferring more energy than was inputted into it. This would break the law of conservation of energy.

3. Efficiency = $\frac{6}{15} = 0.4$

4. Efficiency = $\frac{240}{300} = 0.8$

5. Total input energy transfer = $\frac{5500}{0.80} = 6875\,J$

6. By wrapping the kettle in insulation

Page 209

1. *Any three from:* fossil fuels (coal, oil and natural gas); nuclear fuel; biofuel; wind; hydroelectric; geothermal; tides; the Sun; water waves

2. *Any two from:* transport; generating electricity; heating

3. Coal; oil; (natural) gas

4. Non-renewable energy resources are those that cannot be replaced within a lifetime and will eventually run out.

5. Advantage: *Any one from:* coal is relatively cheap and easy to obtain; coal-fired power stations can be set up relatively quickly
Disadvantage: *Any one from:* burning coal produces carbon dioxide, a greenhouse gas; burning coal produces more carbon dioxide per unit of energy than oil or gas does; burning coal produces sulfur dioxide, which causes acid rain and is costly to remove from the process

6. A renewable energy resource is one that can be replenished as it is used. Examples include: biofuel; wind; hydroelectric; geothermal; tides; Sun; water waves

7. Advantage: *Any one from:* no fuel and little maintenance required for wind turbines; turbines can be built offshore; no pollutant gases produced
Disadvantage: *Any one from:* wind turbines can cause noise and visual pollution; wind is not very flexible in meeting energy demand; large capital outlay

Page 211

1. A complete path around which an electric current can flow.

2. A cell or a battery of cells
At least one device that can transfer energy, e.g. a lamp
Wires, called connecting leads, which join the cell to the components in the circuit
A switch is usually included

3. a)   b)

   c)   d)

4.

5.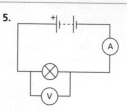

Page 213

1. The rate of flow of electric charge

2. Charge = current × time

3. $Q = 3 \times 600 = 1800$ C

4. $Q = 15 \times 480 = 7200$ C

5. $t = \frac{2000}{1.5} = 1333.3$ s

6. $I = \frac{1200}{900} = 1.3$ A

Page 215

1. The electric current will increase.

2. Potential difference = current × resistance

3. $V = 0.005 \times 300 = 1.5$ V

4. $V = 3.0 \times 250 = 750$ V

5. $I = \frac{12}{4} = 3$ A

6. $R = \frac{400}{10} = 40 \, \Omega$

Page 217

1. Resistance is a measure of how hard it is to get a current through a component at a particular potential difference.
$$\text{Resistance} = \frac{\text{Potential difference}}{\text{current}}$$

2. If the values are directly proportional to one another, when plotted on a graph they will form a straight line with a positive gradient that passes through the origin.

3. a) a length of wire
   b) *Any one from:* LDR; thermistor; filament lamp; diode

4.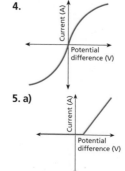

5. a)

b) The curve is this shape because no current flows through the diode in the reverse direction, making the resistance very high. Current only flows in one direction.

6. a) *Any suitable use, e.g.* thermostats in heating systems

b) *Any suitable use, e.g.* switching lights on when it gets dark

Page 219

1. 0.3 A

2. As potential difference is shared between components, the value of the potential difference across each bulb will be $\frac{4.5}{2} = 2.25\,V$

3. They are connected in separate loops.

4. Ammeter X will have the highest reading. This is because the total current in the circuit (measured by ammeter X) is equal to the sum of the currents flowing through the individual components (measured by ammeters Y and Z).

5. *Table completed as follows:*
Series circuit column (from top): same; shared; sum
Parallel circuit (from top): sum; same; less

Page 221

1. An alternating potential difference produces an electric current that regularly changes direction.

2.

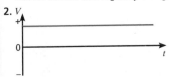

3. Frequency = 50 Hz
Potential difference = 230 V

4. The mains supply is an alternating supply. This allows the electricity to be transmitted more efficiently.

5. Copper is used because it is a good conductor of electricity.

6.

Wire	Function
Live	Carries **alternating** current from the **supply** to the **device**.
Neutral	**Completes** the circuit to allow **current** to flow to the device.
Earth	**Safety** wire; this stops the appliance from becoming **live**.

Page 223

1. Electrical power is a measure of the rate of energy transfer. The greater the power, the faster the rate of energy transfer.

2. $P = 0.7 \times 230 = 161\,W$

3. $P = 8.45^2 \times 35 = 2499\,W$

4. $I = \frac{8000}{250} = 32\,A$

5. $V = \frac{4.5}{3} = 1.5\,V$

6. $R = \frac{2600}{13^2} = 15.4\,\Omega$ (to 3 s.f.)

7. $I = \sqrt{\frac{900}{100}} = 3\,A$

Page 225

1. The power of the appliance.
The time it is switched on for.

2. Energy transferred = power × time

3. $E = 60 \times (20 \times 60) = 72000\,J$

4. $t = \frac{180000}{1500}$
$t = 120$ seconds = 2 minutes

5. $P = \frac{702000}{(3 \times 60 \times 60)} = 65\,W$

6. Energy = charge × potential difference

7. $E = 230 \times 6000 = 1380000\,J$

8. $Q = \frac{165600}{230} = 720\,C$

9. $V = \frac{150}{100} = 1.5\,V$

Page 227

1. C

2. Step-up transformers are used to increase the potential difference so that electricity is transmitted at a high voltage.

3. If electricity is transmitted at a very high voltage, the current will be very low. This reduces the energy loss through heating, making the electricity transmission more efficient.

4. Step-down transformers are used to decrease the potential difference so that electricity can be delivered to consumers at a safe voltage.

5. a) $\frac{V_p}{V_s} = \frac{N_p}{N_s}$
$\frac{230}{1000} = \frac{50}{N_s}$
$N_s = 217$ turns

b) Step-up (as the potential difference is increased from primary to secondary)

6. $V_p I_p = V_s I_s$
So $I_p = \frac{(V_s I_s)}{V_p}$
$I_p = \frac{(230 \times 13)}{400000}$
$I_p = 7.5 \times 10^{-3}\,A$

Page 229

1. mass and volume

2. Since the volume of a gas is larger than the volume for the same mass as a solid, the density of a gas is less than that of the solid.

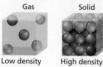

Gas — Low density   Solid — High density

3. Density = $\dfrac{\text{mass}}{\text{volume}}$

Density is measured in kg/m³

Mass is measured in kg

Volume is measured in m³

4. $\rho = \dfrac{1134.3}{0.1} = 11\,343\,\text{kg/m}^3$

5. $m = \dfrac{7.5}{1000} = 7.5 \times 10^{-3}\,\text{kg}$

$v = \dfrac{7.5 \times 10^{-3}}{19\,320}$

$v = 3.88 \times 10^{-7}\,\text{m}^3$

6. Since the density of the iron is almost 8 times that of the water, the iron nail will sink in water.

Page 231

1.

Sublimation, Melting, Boiling, Solid, Liquid, Gas, Freezing, Condensation

2. a) The temperature at which a pure substance changes from a solid to a liquid.

b) The temperature at which a pure substance changes from a liquid to a gas.

3.

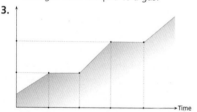

Time

4. Similarity: Both evaporation and boiling are physical changes that involve a liquid turning into a gas.

Difference *(any one from)*: Evaporation happens at any temperature between the melting point and boiling point of a substance, whereas boiling only happens at the boiling point; The energy for evaporation comes mainly from the kinetic energy of the substance, whereas the energy for boiling comes mainly from the surroundings of the substance.

5. The internal energy of a system is the sum of the kinetic and potential energies of all of the particles that make up the system.

Page 233

1. The mass of the substance being heated.

The substance being heated.

The energy input.

2. The energy required to increase the temperature of 1 kg of a substance by 1°C.

3. The substance with the **lower** specific heat capacity will have a greater temperature rise.

4. change in thermal energy = mass × specific heat capacity × change in temperature

5. $\Delta E = (5 \times 10^{-3}) \times 2000 \times 80$

$\Delta E = 800\,\text{J}$

6. $c = \dfrac{146\,250}{(0.25 \times 30)} = 19\,500\,\text{J/kg°C}$

7. $m = \dfrac{101\,500}{(900 \times 205)}$

$m = 0.55\,\text{kg} = 550\,\text{g}$

Page 235

1. The energy that is added to a substance as it changes state. This is the energy needed to overcome the intermolecular forces.

2. The specific latent heat of fusion is the energy required to melt 1 kg of a substance, whereas the specific latent heat of vaporisation is the energy required to change the state of 1 kg of a substance from liquid to gas.

3. energy = mass × specific latent heat

4. $E = \left(\dfrac{20}{1000}\right) \times 275\,700 = 5514\,\text{J}$

5. $L = \dfrac{1\,995\,000}{5} = 399\,000\,\text{J/kg}$

6. $m = \dfrac{1\,710\,000}{855\,000} = 2\,\text{kg}$

Page 237

1. *Any two from:* particles are identical unless they are given different colours; the individual atoms in the particles are not shown; intermolecular forces between particles are not shown

2. The particles in a gas move with a constant, random motion (in random directions) at a range of different speeds.

3. An increase in temperature leads to an increase in the average kinetic energy of the particles.

4. $E_k = \dfrac{1}{2}mv^2$

As kinetic energy is proportional to the square of the speed, if the average speed doubles, the kinetic energy quadruples.

5. Pressure is caused by collisions between the gas particles and the walls of the container in which it is held.

6. If the temperature is increased, the average kinetic energy of the particles increases. This means that they will collide with the walls of the container more frequently and with more energy. The increasing frequency of collisions raises the pressure of the gas, since each collision creates a force.

Page 239

1.

Atomic particle	Relative mass	Relative charge
Proton	1	+1
Neutron	1	0
Electron	$\frac{1}{2000}$	−1

2. Ratio $= \frac{1.5 \times 10^{-14}}{1.7 \times 10^{-15}} = 8.8$

8.8 times bigger

3. Number of neutrons $= 27 - 13$
Number of neutrons $= 14$

4. Number of protons $= 12$
Number of electrons $= 12$
Number of neutrons $= 12$

5. Isotopes are atoms of the same element with the same number of protons, but a different number of neutrons.

6. Similarity: all three isotopes have 1 proton and 1 electron
Difference: protium has no neutrons, deuterium has 1 neutron and tritium has 2 neutrons.

Page 241

1. Atoms as tiny, solid, indivisible spheres.

2. Electron

3.

4.

Most alpha particles passed straight through the gold foil		The nucleus is very small and very dense
Some alpha particles were deflected at small angles		Most of the atom is empty space
A very small number of alpha particles were deflected straight back		The nucleus is positively charged

5.

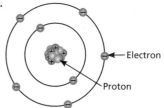

Electron

Proton

6. James Chadwick

Page 243

1. The nuclei of unstable atoms may disintegrate and emit radiation.

2. It is impossible to tell which nuclei will decay or when a particular nucleus will decay.

3. A helium nucleus (or two protons and two neutrons)

4. An alpha particle contains two protons so when a nucleus emits an alpha particle, the proton number of the remaining nucleus decreases by two. Since the proton number determines the element, a different proton number means a new element has been formed.

5. proton and electron

6. Ionisation can damage or kill healthy cells within the body.

7. The activity is the rate at which unstable nuclei decay. The count rate is the rate at which the decay is detected.

8.

Type of radiation	Penetration power	Range of air	Ionising power
Alpha	Stopped by **skin / a sheet of paper**	< 5 cm	**High**
Beta	Stopped by 2–3 mm of aluminium foil	~ 1 m	Low
Gamma	Stopped by **thick lead / concrete**	> 1 km	**Very low**

Page 245

1. a) $^{4}_{2}\text{He}$ b) $^{0}_{-1}\text{e}$

2. Gamma radiation is in the form of an uncharged electromagnetic wave, which has no mass, and so the emission of a gamma wave does not change the charge or mass of the nucleus.

3. *Table completed as follows:* Decreases by **2**; **Decreases** by 4

4. *Table completed as follows:* Increases by **1**;
Remains unchanged

5. $X = 235$ $Y = 94$

6. *Missing values are (from the left):* 11; 7; 0

Page 247

1. It is not possible to tell which nuclei in a sample will decay, or when a particular nucleus will decay.

2. *Any one from:* The time taken for the number of nuclei in a sample to halve; The time taken for the activity / count rate of the sample to fall to half of its initial value

3. 5 minutes

4.

Time (days)	Activity (Bq)
0	600
5	300
10	150
15	75
20	37.5
25	18.75
30	9.375

Activity after 30 days will be 9.375 Bq

5. a) $\frac{120}{30} = 4$ half-lives

$(\frac{1}{2})^4 = \frac{1}{16}$

So, $\frac{1}{16}$th will remain after 120 years.

b) $1000 \times (\frac{1}{16}) = 62.5$ g

Page 249

1. *Any suitable similarity and difference, e.g.*
Similarity: Both irradiation and contamination can cause damage to living cells and so must be carefully considered.
Difference: *(any one from)*: Irradiation is the exposure of an object to radiation, whereas contamination is the presence of material containing radioactive atoms on or in the object; If an object is irradiated, this stops as soon as the source is removed and the object does not become radioactive, whereas a contaminated object will remain radioactive for as long as the source is on or in it.

2. The food is exposed to a gamma source. The gamma rays destroy living tissue and therefore kill bacteria on the surface of the food.

3. *Any suitable use, e.g.* sterilising surgical instruments; destroying cancerous tumours

4. a) A radioactive source could be introduced into the gas supply. The movement of this source could then be tracked. At the site of the leak, large amounts of radiation would be detected.

b) The half-life of the source should be carefully considered. It needs to be long enough for the detection to occur, but short enough that the source is not active for longer than is necessary to detect the leak. For this reason, a source with a half-life of hours, rather than days, should be used.

Page 251

1. A scalar quantity is a quantity with only magnitude (size)

2. *Any two from:* energy; mass; time; temperature; speed

3. A scalar quantity has only magnitude whereas a vector quantity has magnitude and direction.

4. *Any two from:* force; velocity; momentum; acceleration; displacement

5.

$$\overset{40\,\text{N}}{\longleftarrow} \blacksquare \overset{70\,\text{N}}{\longrightarrow}$$

The resultant force is 30 N to the right.

6. *Any three from:* friction; air resistance; normal contact force; tension

7. gravitational; electrostatic; magnetic

Page 253

1. The gravitational field strength is the gravitational force on a 1 kg object. It is measured in N/kg.

2.

3. Weight is the force of gravity on an object. It is measured in newtons, N.

4. Weight = mass × gravitational field strength

5. $W = 5 \times 10 = 50$ N

6. $\frac{W}{m} = g$

$g = \frac{104}{65} = 1.6$ N/kg

Page 255

1. Resultant force = 10 − 5
Resultant force = 5 N to the right

2. Resultant force = 7 − 5
Resultant force = 2 N downwards

3.

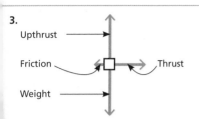

Labels: Upthrust, Friction, Thrust, Weight

4.

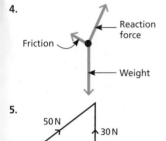

Labels: Reaction force, Friction, Weight

5.

50 N, 30 N, 31°, 40 N

Horizontal component = 40 N

Vertical component = 30 N

Page 257

1. When a force moves an object, work is done.

2. Work done = energy transferred

3. Work done = force × distance. Work done is measured in joules (J), force is measured in newtons (N) and distance is measured in metres (m).

4. $W = 15 \times 2 = 30\,J$

5. $W = Fs$, so $F = \dfrac{W}{s}$

$F = \dfrac{240}{12} = 20\,N$

6. $W = Fs$, so $s = \dfrac{W}{F}$

$F = \dfrac{4500}{1000} = 4.5\,m$

Page 259

1. Extension = 60 − 20 = 40 cm

Extension = 0.4 m

2. Compression = 5 − 1 = 4 cm

3. Force = spring constant × extension

Force is measured in newtons (N); spring constant is measured in newtons per metre (N/m); extension is measured in metres (m)

4. The graph is a straight line that passes through the origin. This shows that extension and force are directly proportional.

5. $E_e = \frac{1}{2}ke^2$

$E_e = \frac{1}{2} \times 32 \times 0.3^2 = 1.44\,J$

6. $\sqrt{\dfrac{2E_e}{k}} = e$

$e = \sqrt{\dfrac{2 \times 20}{2.5}} = 4\,m$

7. $\sqrt{\dfrac{2E_e}{k}} = e$

$e = \sqrt{\dfrac{2 \times 15}{50}} = 0.77\,m$

Final length = 0.05 + 0.77 = 0.82 m

So 82 cm

Page 261

1. Distance is the total number of metres, kilometres or miles travelled, whereas displacement is the distance travelled in a straight line in a given direction from start to finish.

2. Distance travelled and time taken

3. *Any suitable answers, e.g.*

Walking	1.5 m/s
Running	3 m/s
Cycling	6 m/s
Sound in air	330 m/s

4. Distance = speed × time

Distance is measured in metres (m); speed is measured in metres per second (m/s); time is measured in seconds (s)

5. $s = 3.5 \times 1800$

$s = 6300\,m = 6.3\,km$

6. $v = \dfrac{s}{t}$

$v = \dfrac{21\,600}{3600} = 6\,m/s$

7. Velocity is the speed of an object in a given direction. Even if the speed is constant, an object in a circle is constantly changing direction, which means its velocity is also constantly changing.

Page 263

1. Acceleration is the rate of change of velocity. It is measured in metres per second squared (m/s²).

2. Acceleration = $\dfrac{\text{change in velocity}}{\text{time}}$

3. $a = \dfrac{\Delta v}{t}$

$a = \dfrac{6}{6} = 1\,m/s^2$

4. $a = \dfrac{\Delta v}{t}$

$\Delta v = a \times t$

$\Delta v = 2 \times 15 = 30\,m/s$

5. a) The acceleration of the object

b) The total distance travelled by the object

6. a)

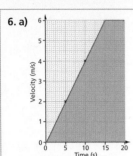

b) Finding the gradient of the line over the first 15 seconds = $\frac{6}{15}$ = 0.4 m/s^2

c) Calculating the area under the line =
$(0.5 \times 15 \times 6) + (5 \times 6) = 75$ m

Page 265

1. An object will remain stationary or move at a constant speed, in the same direction, unless acted upon by a resultant force.

2. The forces of weight and normal reaction force are equal and opposite. This means that there is no resultant force acting on the book. According to Newton's first law, no resultant force means that a stationary object will have zero acceleration.

3. An object will accelerate when acted upon by a resultant force.

4. $F = 20\,000 \times 2 = 40\,000$ N

5. $a = \frac{F}{m} = \frac{500}{100} = 5$ m/s^2

6. For every action force, there is an equal and opposite reaction force.

7. The pair of forces must:
- act on different bodies
- act in opposite directions
- act along the same line
- be of the same type (e.g. both gravitational forces)
- be of the same magnitude.

Page 267

1. 0.2 to 0.9 s

2. a) The distance travelled during the driver's reaction time.

b) The distance travelled between the driver applying the brakes and the car coming to a stop.

3. $9 + 14 = 23$ m

4. *Any three from:* Tiredness – if a driver is tired, their reaction time will increase; Taking drugs – if a driver is taking medication or illegal substances, it may increase reaction time; Drinking alcohol – this also increases reaction time; Distractions, such as using a mobile phone or children playing in the back of a car, will increase reaction time.

5. The speed of the vehicle – greater speed increases the braking distance; The road conditions – wet or icy roads can reduce the friction between the tyres and the road surface, increasing the braking distance; The condition of the vehicle – worn brakes or worn tyres increase the braking distance.

6. The increase in frictional forces between the brakes and the wheels can cause the brakes to overheat.

Page 269

1. Momentum = mass × velocity

2. $p = 80 \times 3 = 240$ kg m/s

3. $v = \frac{15\,000}{1500} = 10$ m/s

4. $p = mv$
$m = \frac{p}{v}$
$m = 67\,500 \div 9 = 7\,500$ kg

5. Momentum of bullet = $0.02 \times 500 = 10$ kg/s
Momentum of cheetah = $50 \times 30 = 1500$ kg/s
The cheetah has the greater momentum.

6. In a closed system, the momentum before a collision is equal to the momentum after a collision.

7. As the moving toy car hits the stationary car, both cars will move off in the same direction as the moving car was travelling, but at a lower speed. Momentum must be conserved before and after the collision, so the direction must be in the same direction as the moving car, and the velocity lower as the mass increases as the cars join together.

Page 271

1. a) *One of:* light; water
b) Sound

2.

Crest — Direction of travel of the wave

Trough

3. 35°

4. The amplitude of a wave is the maximum displacement of the wave from its undisturbed position.

5. The frequency of a wave is the number of waves passing a given point every second. The unit of frequency is hertz (Hz).

6. $f = \frac{1}{T} = \frac{1}{10}$
$f = 0.1$ Hz

7. Wave speed = frequency × wavelength

8. $v = f\lambda = 500 \times 0.001$
   $v = 0.5\,\text{m/s}$

9. $v = f\lambda$
   So $\lambda = \frac{v}{f}$
   $\lambda = \frac{(3 \times 10^6)}{3000} = 1000\,\text{m}$

Page 273

1. radio waves, microwaves, infrared, visible light, ultraviolet, X-rays, gamma rays

2. Speed = $(7.0 \times 10^{14}) \times (4.3 \times 10^{-7})$
   Speed = $3.0 \times 10^8\,\text{m/s}$

3. Refraction is the change of direction of a wave as it crosses a boundary between two transparent materials.

4. UV radiation; X-rays; gamma rays

5. If radio waves are absorbed, they may create an alternating current with the same frequency of the radio waves.

6. X-rays and gamma rays are both types of ionising radiation. This means that they can cause mutations in cells in the body, which can cause cancer.

Page 275

1.

Radio waves	Remote controls for televisions
Microwaves	Transmitting TV signals
Infrared radiation	Producing shadow pictures of bones
Ultraviolet radiation	Satellite communications
X-rays	Killing bacteria on food
Gamma rays	Security coding

2. The microwaves can be absorbed by water in the cells in the body, which can heat up, causing them to become damaged.

3. X-rays and gamma rays pass through soft tissue (some is absorbed). A high dose of X-rays or gamma rays can kill cells in the body; a low dose can lead to cancerous tumours forming in the body.

4. Ultraviolet (UV) radiation passes through the skin to the tissues below. A high dose of UV radiation can kill body cells; a low dose of UV radiation can lead to cancer.

Page 277

1. North pole; south pole

2. Two bar magnets can be placed close to one another, but not touching. If the two north poles are close, the magnets will repel. If the north pole of one magnet is placed close to the south pole of the second magnet, they will attract each other.

3. A region around a magnet where a force acts on another magnet or magnetic material

4. *Any two from:* nickel; steel; cobalt; neodymium

5. A permanent magnet produces its own magnetic field.
   Any suitable example, e.g. iron; cobalt; nickel

6. Repeatedly stroking the nail with the pole of a magnet from head to point magnetises the nail.

7. A material that becomes magnetic when placed in a magnetic field

Page 279

1. A coil of wire with an electric current passing through it. When the current flows, a magnetic field is generated around the coil of wire.

2. A long coil of wire with a soft iron core

3.

Magnet	Type of magnet	Strength of field
Bar magnet	**Permanent**	Quite strong
Electromagnet with air core	Temporary	**Weak**
Solenoid with iron core	**Temporary**	Very strong

4. Electromagnets can be switched on or off by disconnecting the battery.
   The current of an electromagnet can be changed, so the strength of the magnetic field can be controlled.
   The direction of the current in an electromagnet can be reversed by reversing the battery connections.

5. *Any three from:* loudspeakers; magnetic locks; data storage; headphones; cranes; electric bells

Page 281

1. When a wire (conductor) carrying an electric current is placed in a magnetic field, the magnetic field formed around the wire interacts with the permanent magnetic field and this causes the wire to experience a force, which makes it move.

2. By increasing the size of the current
By using stronger magnets / increasing the strength of the magnetic field

3. By turning the cells around / reversing the direction of flow of current
By reversing the direction of the magnetic field

4. Lining up the first finger with the magnetic field, pointing left to right (from north to south), and the second finger with the current, pointing down (from positive to negative), the thumb will point towards us. Therefore, the wire will move in this direction.

5. Force = magnetic flux density × current × length

6. $F = BIL = 1.5 \times 10 \times 2.5$
$F = 37.5\,N$

Page 283

1. An electromagnet / a wire with a current flowing through it

2. A battery is connected to the coil, which becomes an electromagnet. The coil's magnetic field and the field produced by the magnets interact. The interaction creates an upward force on one side of the coil. The interaction also creates a downward force on the other side of the coil. The upward and downward forces make the coil rotate clockwise.

3. *Any three suitable answers, e.g.* washing machine; tumble dryer; microwave oven; electric tin opener; electric food mixer; electric fan

4. If the magnetic flux density doubles, the force also doubles (as force is directly proportional to the magnetic flux density).

5. a) $F = BIL$
$F = 2.5 \times 5 \times 2$
$F = 25\,N$ (to 2 s.f.)
b) $F = BIL$
$F = 2.5 \times 2.5 \times 1.5$
$F = 13\,N$ (to 2 s.f.)

Mixed questions
Pages 284–289 Paper 1
Biology

1. a) Digestion of lipids produces fatty acids. As they are acids, they have a low pH.
b) The lipase has been boiled so it has been denatured and will not work as a catalyst.

2. A cell that is hollow and forms tubes – To transport water
A cell that has a flagellum – To swim
A cell that is full of protein fibres – To contract
A cell that has a long projection with branched endings – To carry nerve impulses

3. lungs; vein; aorta; deoxygenated; vena cava

4. a) Xylem **b)** Phloem

5. glucose + oxygen → carbon dioxide + water

6.

Type of cell	Nucleus	Cytoplasm	Cell membrane	Cell wall
Plant cell	✓	✓	✓	✓
Bacterial cell	✗	✓	✓	✓
Animal cell	✓	✓	✓	✗

7. a) Osmosis involves movement of water molecules whereas diffusion is movement of particles. Osmosis always involves a membrane whereas diffusion may not.
b) Active transport is movement of particles against a concentration gradient / from low to high concentration, rather than from high to low concentration as with diffusion. Active transport requires energy from respiration whereas diffusion does not.

Chemistry

8. Halogens

9.

10. a) covalent
b) 2,8,8
c) An electrostatic force of attraction between oppositely charged ions.

11. When melting hydrogen chloride, no bonds are broken. There are only weak forces of attraction between molecules, but when sodium chloride is melted, many strong bonds must be broken. It requires more energy to break bonds than overcome forces of attraction.

12. a) $(2 \times 14) + (4 \times 1) + (3 \times 16) = 80$
b) $\left(\frac{(2 \times 14)}{80}\right) \times 100 = 35\%$

13. a) Oxidation is gain of oxygen (*Accept* loss of electrons).
b) magnesium + oxygen → magnesium oxide

14. a) Magnesium
b) Copper sulfate
c) copper sulfate + magnesium → magnesium sulfate + copper

15.

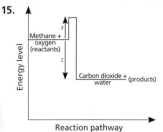

(Reactants higher in energy than products. Activation energy marked (x), energy given out marked (z).)

Physics

16. $\dfrac{2E_k}{v^2} = m$

$m = \dfrac{(2 \times 203)}{90^2}$

$m = 0.05\,kg = 50\,g$

17. $E_p = mgh$

$m = 0.25\,kg$

$E_p = 0.25 \times 9.8 \times 2.5$

$E_p = 6.125\,J = 6.1\,J$ (to 2 s.f.)

18. $P = I^2R = 5^2 \times 100$

$P = 2500\,W$

19. *Any suitable advantages, e.g.* fast start-up time; no pollutant gases emitted; water can be pumped back to the reservoir when electricity demand is low

Any suitable disadvantages, e.g. only certain locations are suitable (it often involves damming upland valleys); there must be adequate rainfall in the region where the reservoir is; very high initial capital outlay

20. $Q = It$

$I = \dfrac{Q}{t} = \dfrac{4500}{(15 \times 60)}$

$I = 5\,A$

21. $V = IR$

$\dfrac{V}{I} = R$

$R = \dfrac{230}{(0.1)} = 2300\,\Omega$

22.

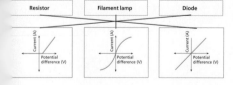

23. $E = Pt$

$t = \dfrac{E}{P}$

$t = \dfrac{135\,000}{900}$

$t = 150\,s$

24. $\underset{53}{131}\,I \;\rightarrow\; \underset{54}{131}\,Xe \;+\; \underset{-1}{0}\,e$

25. $\dfrac{\Delta E}{(mc)} = \Delta\theta$

$\Delta\theta = \dfrac{450\,000}{(2.0 \times 840)}$

$\Delta\theta = 270°C$

Pages 290–295 Paper 2
Biology

1. a) grass / lettuce → caterpillar → thrush → hawk

b)

Type of organism	Number in the diagram
Producers	2
Herbivores	2
Prey	3
Apex predators	1

2. male gamete = 23, zygote = 46, embryo = 46

3. This is a reflex that is not under conscious control. There are no neurones from the brain / relay neurones preventing the motor neurone from making the muscle contract.

4. a) *Equus*

b) Two animals in the same species cannot have two different scientific names as that is the name of the species. Animals often have two different common names.

5. Choose two zebras that have the fewest stripes on their backs. Mate these zebras together and then choose the offspring that have the fewest stripes. Mate these together and repeat the process.

6. genetic engineering

Chemistry

7. The minimum energy needed to start a reaction.

8. a) Concentration of acid

b) g

c) In a higher concentration there will be more acid particles in a given volume. This means that there will be more collisions in the same amount of time, so more successful collisions in a given time and therefore a faster rate of reaction.

9. a) Ethane

   b) C_nH_{2n+2}

10. a) Ca^{2+}

   b) The white precipitate formed doesn't dissolve with excess sodium hydroxide solution.

11. a) Nitrogen

   b) O_2

12. 200 million years

13. Soot / carbon

14. a) A molecule that contains only hydrogen and carbon

   b) Cracking

Physics

15. $\dfrac{W}{m} = g$

$g = \dfrac{112}{70} = 1.6\,\text{N/kg}$

16. $W = Fs$, so $s = \dfrac{W}{F}$

$F = \dfrac{500}{50} = 10\,\text{m}$

17. $v = f\lambda$

So $\lambda = \dfrac{v}{f}$

$\lambda = \dfrac{(3 \times 10^8)}{(4 \times 10^{14})} = 7.5 \times 10^{-7}\,\text{m}$

18. $s = vt$

$v = \dfrac{s}{t}$

$v = \dfrac{15\,500}{1800} = 8.6\,\text{m/s}$

19. $a = \dfrac{(v - u)}{t}$

$u = v - at$

$u = 10 - (2 \times 3) = 4\,\text{m/s}$

20. $v = \dfrac{p}{m}$

$v = \dfrac{11\,250}{2500} = 4.5\,\text{m/s}$

21. $\sqrt{\dfrac{2Ee}{k}} = e$

$e = \sqrt{\dfrac{(2 \times 40)}{4.5}} = 4.2\,\text{m}$

22. a) Radio waves

   b) Microwaves

   c) Infrared radiation

23. North to South

24. To increase the strength of the magnetic field around the coil of wire

25. *Any one from:* move the magnet out of the coil; move the other pole of the magnet into the coil

26. a) $\dfrac{V_p}{V_s} = \dfrac{N_p}{N_s}$

$\dfrac{230}{4500} = \dfrac{100}{N_s}$

$N_s = 1957$ turns

   b) Step-up transformer as the potential difference is increased from the primary coil to secondary coil, and the number of turns is greater on the secondary coil than the primary coil.

27. $F = BIL$

$F = 0.75 \times 13 \times 0.5$

$F = 4.9\,\text{N (to 2 s.f.)}$